Praise for *Quick and Nimble*

"Adam Bryant gives us yet another welcome opportunity to listen in on his living laboratory of leadership conversations. In *Quick and Nimble*, he orchestrates discussion on the idea—and its practical application—that culture is not in 'support' of strategy; it *is* strategy. Bryant's gift for asking incisive questions of remarkable people, and translating their insights into crisp and lucid prose, makes for joyful reading indeed—smart, provocative, and immensely useful!"
—Jim Collins, author of *Good to Great*
and coauthor of *Great by Choice*

"Adam Bryant identifies the most important challenge facing business leaders today: how to achieve innovation at scale by building a culture that will marry the energy of a start-up with the discpline of a veteran organization. In *Quick and Nimble*, he has assembled an all-star team of CEOs to share—in their own words—how they have taken their companies to the next level."
—Eric Ries, bestselling author of *The Lean Startup*

"In today's nonstop turbulence, innovation is critical to success. In *Quick and Nimble*, Adam Bryant distills important lessons from a range of leading CEOs about how to create and sustain a productive culture that nurtures not only innovation but also respect, engagement, and teamwork for everyone involved. And he does so in a lively, grounded voice that resonates with experience and perspective. A 'must-read' for any leader serious about the company he or she is building and the difference it makes in the world."

—Nancy F. Koehn, James E. Robison
Professor of Business Administration, Harvard Business School

"In *Quick and Nimble*, Adam Bryant unlocks the secret to creating and sustaining a culture of innovation: Leadership. Bryant has translated his in-depth interviews with innovation leaders into lessons on how to build a creative, open, and ultimately innovative culture that will enable every organization to thrive and grow."
—Bill George, professor, Harvard Business School, and former chair and CEO of Medtronic

"Compelling . . . [*Quick and Nimble*] offers a realistic plan for achieving extraordinary results. . . . Bryant presents the perfect combination of overarching philosophy and nuts-and-bolts advice."
—*Daily News* (Bowling Green, KY)

"A powerful pick for any business leader or entrepreneur."
—*The Midwest Book Review*

"Reams of practical advice for and from business leaders most—thankfully—with a human, caring touch."   —*Kirkus Reviews*

"Leaders such as the editor in chief of *Teen Vogue,* the CEO of JetBlue, chef Mario Batali . . . provide thoughts on leadership, management, and innovation . . . The big names lend the book a certain flair . . . [Bryant] breaks down the important elements of a successful corporate culture, and then describes the leadership strategies to build on it."   —*Publishers Weekly*

"Illuminating . . . The author's natural, conversational tone is compulsively readable."   —*Library Journal*

Also by Adam Bryant

The Corner Office:
Indispensable and Unexpected Lessons
from CEOs on How to Lead and Succeed

QUICK AND NIMBLE

LESSONS FROM LEADING
CEOS ON HOW TO CREATE
A CULTURE OF INNOVATION

Adam Bryant

St. Martin's Griffin
New York

www.stmartins.com

Designed by Meryl Sussman Levavi

The Library of Congress has cataloged the Times Books edition as follows:

Bryant, Adam.
  Quick and nimble : lessons from leading CEOs on how to create a culture of innovation / Adam Bryant.—First edition.
    p. cm.
  Includes index.
  ISBN 978-0-8050-9701-6 (hardcover)
  ISBN 978-0-8050-9702-3 (e-book)
  1. Corporate culture. 2. Creative ability in business. 3. Diffusion of innovations—Management. 4. Technological innovations. 5. New products. I. Title.
  HD58.7.B797 2014
  658.4—dc23

2013014348

ISBN 978-1-250-06084-6 (trade paperback)

St. Martin's Griffin books may be purchased for educational, business, or promotional use. For information on bulk purchases, please contact the Macmillan Corporate and Premium Sales Department at 1-800-221-7945, extension 5442, or write to specialmarkets@macmillan.com.

First published in hardcover by Times Books, an imprint of Henry Holt and Company, LLC

First St. Martin's Griffin Edition: January 2015

10 9 8 7 6 5 4 3 2 1

To my parents

CONTENTS

   8. Play It Again and Again   107

   9. Building Better Managers   115

   10. Surfacing Problems   132

   11. School Never Ends   147

   12. The Art of Smarter Meetings   158

   13. Knocking Down Silos   173

   14. Sparking Innovation   187

   15. Can We Have Some Fun?   221

   16. Alone at the Top   230

      Conclusion   243

      Acknowledgments   249

      Index   253

LIST OF INTERVIEWS

Jeremy Allaire, CEO, Brightcove

Nancy Aossey, CEO, International Medical Corps

Shellye Archambeau, CEO, MetricStream

Amy Astley, editor in chief, *Teen Vogue*

Joel Babbit, CEO, Mother Nature Network

Romil Bahl, CEO, PRGX

Chris Barbin, CEO, Appirio

David Barger, CEO, JetBlue

Carl Bass, CEO, Autodesk

Mario Batali, chef

Charlotte Beers, former CEO, Ogilvy & Mather Worldwide

Seth Besmertnik, CEO, Conductor

Lars Björk, CEO, QlikTech

Lynn Blodgett, CEO, ACS

Laszlo Bock, senior vice president for people operations, Google

Bob Brennan, former CEO, Iron Mountain

Linda Lausell Bryant, executive director, Inwood House

Tim Bucher, former CEO, TastingRoom.com

Richard R. Buery Jr., CEO, Children's Aid Society

Geoffrey Canada, CEO, Harlem Children's Zone

Steve Case, CEO, Revolution

Annette Catino, CEO, QualCare

Laura Ching, chief merchandising officer, TinyPrints.com

Cathy Choi, president, Bulbrite

Andrew Cosslett, former CEO, Intercontinental Hotels Group

Susan Credle, chief creative officer, Leo Burnett USA

Dennis Crowley, CEO, Foursquare

Chris Cunningham, CEO, Appssavvy

Barbara DeBuono, CEO, Orbis International

Tracy Dolgin, CEO, YES Network

Robin Domeniconi, former chief brand officer, Elle Group

John Donahoe, CEO, eBay

John Donovan, senior executive vice president of AT&T Technology and Network Operations

John Duffy, CEO, 3Cinteractive

Kris Duggan, chief strategy officer, Badgeville

Liz Elting, co-CEO, TransPerfect

Richard D. Fain, CEO, Royal Caribbean Cruises

Deborah Farrington, general partner, StarVest Partners

Kenneth Feld, CEO, Feld Entertainment

Pamela Fields, CEO, Stetson

Kathleen Flanagan, CEO, Abt Associates

Bill Flemming, president, Skanska USA Building

Christine Fruechte, CEO, Colle + McVoy

Mark Fuller, CEO, WET Design

Russell Goldsmith, CEO, City National Bank

Ilene Gordon, CEO, Ingredion

William D. Green, former CEO, Accenture

Julie Greenwald, COO, Atlantic Records Group

F. Mark Gumz, former CEO, Olympus Corporation of
  the Americas

Amy Gutmann, president, University of Pennsylvania

Jen Guzman, CEO, Stella & Chewy's

Ori Hadomi, CEO, Mazor Robotics

Katherine Hays, CEO, GenArts

Linda Heasley, former CEO, The Limited

Daniel T. Hendrix, CEO, Interface

Angie Hicks, chief marketing officer, Angie's List

Tony Hsieh, CEO, Zappos.com

Jen-Hsun Huang, CEO, Nvidia

Joseph Jimenez, CEO, Novartis

Robert L. Johnson, chairman, RLJ Companies

Lily Kanter, cofounder, Serena & Lily

Marjorie Kaplan, president, Animal Planet and Science cable
  networks

Jeffrey Katzenberg, CEO, DreamWorks Animation

Ivar Kroghrud, lead strategist, QuestBack

Barbara J. Krumsiek, CEO, Calvert Investments

Arkadi Kuhlmann, former CEO, ING Direct

Andy Lansing, CEO, Levy Restaurants

Sir Terry Leahy, former CEO, Tesco

Michael Lebowitz, CEO, Big Spaceship

Niki Leondakis, president and COO, Kimpton Hotels and Restaurants

Dawn Lepore, former CEO, Drugstore.com

Ben Lerer, CEO, Thrillist Media Group

Aaron Levie, CEO, Box

Phil Libin, CEO, Evernote

Selina Lo, CEO, Ruckus Wireless

Robert LoCascio, CEO, LivePerson

Doreen Lorenzo, president, Frog Design

Peter Löscher, CEO, Siemens

Susan Lyne, vice chairman, Gilt Groupe

Gregory B. Maffei, CEO, Liberty Media

Sheila Lirio Marcelo, CEO, Care.com

Paul Maritz, former CEO, VMware

Michael Mathieu, former CEO, YuMe

Tracey Matura, general manager, Smart Car

Karen May, vice president for people development, Google

Chauncey C. Mayfield, CEO, MayfieldGentry Realty Advisors

Andy Mills, president, Medline Industries

Jenny Ming, CEO, Charlotte Russe

Jarrod Moses, CEO, United Entertainment Group

Robert J. Murray, CEO, iProspect

Christopher J. Nassetta, CEO, Hilton Worldwide

Vineet Nayar, CEO, HCL Technologies

John Nottingham, co-president, Nottingham Spirk

David C. Novak, CEO, Yum Brands

Jacqueline Novogratz, CEO, Acumen Fund

Dominic Orr, CEO, Aruba Networks

Dinesh C. Paliwal, CEO, Harman International Industries

Victoria Ransom, CEO, Wildfire

Abbe Raven, CEO, A&E Television Networks

Ken Rees, CEO, Think Finance

John Riccitiello, former CEO, Electronic Arts

Laurel J. Richie, president, WNBA

David Rock, director, NeuroLeadership Institute

Dan Rosensweig, CEO, Chegg

Marcus Ryu, CEO, Guidewire

David Sacks, CEO, Yammer

Stephen I. Sadove, CEO, Saks

Enrique Salem, former CEO, Symantec

Martha S. Samuelson, CEO, Analysis Group

Brent Saunders, former CEO, Bausch & Lomb

Kathy Savitt, former CEO, Lockerz

Dan Schneider, CEO, SIB Development and Consulting

Amy Schulman, executive vice president and general counsel,
  Pfizer

Niraj Shah, CEO, Wayfair.com

Ronald M. Shaich, chairman and co-CEO, Panera Bread

Kevin Sharer, former CEO, Amgen

Mike Sheehan, CEO, Hill Holliday

Irwin D. Simon, CEO, Hain Celestial Group

Ryan Smith, CEO, Qualtrics

John Spirk, co-president, Nottingham Spirk

Caryl M. Stern, CEO, U.S. Fund for UNICEF

Steve Stoute, CEO, Translation LLC; chairman, Carol's Daughter

Tracy Streckenbach, CEO, Hillview Consulting

Shivan S. Subramaniam, CEO, FM Global

Mark B. Templeton, CEO, Citrix

Andrew M. Thompson, CEO, Proteus Digital Health

Terry Tietzen, CEO, Edatanetworks

Kip Tindell, CEO, The Container Store

Tony Tjan, CEO, Cue Ball

Alan Trefler, CEO, Pegasystems

Geoff Vuleta, CEO, Fahrenheit 212

Jeff Weiner, CEO, LinkedIn

Harry West, CEO, Continuum

Jim Whitehurst, CEO, Red Hat

Shawn H. Wilson, president, Usher's New Look Foundation

Catherine Winder, former president, Rainmaker Entertainment

Will Wright, video game designer

Laura Yecies, former CEO, SugarSync

Kyle Zimmer, CEO, First Book

AUTHOR'S NOTE

The material in this book is derived in part from the author's interviews with more than two hundred CEOs and other top executives—though not all are directly quoted in this book—at companies that range in size from small start-ups to large multi-national corporations, as well as nonprofit organizations. The executives' responses were recorded, transcribed, and condensed for publication. Their job titles reflect the status, except where noted, of the executives at the time they were interviewed. The conversations occurred between March 2009 and May 2013.

QUICK AND NIMBLE

INTRODUCTION

We aspire to be the largest small company
in our space.

When Dominic Orr, the CEO of the wireless technology company Aruba Networks, said those words, he crystallized a goal I had heard many leaders express during the more than two hundred interviews I've conducted for the "Corner Office" feature in the *New York Times.* And that idea ultimately helped frame the question that drives this book: How can a company foster a quick and nimble culture—with the enviable qualities of many start-ups—even as it grows?

All leaders and managers face this challenge, regardless of the size of their companies. Even the founders of Google have worried about losing the start-up magic that helped propel the search engine's phenomenal growth. When Larry Page announced in January 2011 that he was taking over the CEO role from Eric Schmidt, he explained to reporters that the company needed to move faster and recapture the agility of its early days, before it grew into a colossus.

"One of the primary goals I have," Page said at the time, "is to get Google to be a big company that has the nimbleness and soul and passion and speed of a start-up."

Discussions of corporate culture can easily fall into platitudes, theories, truisms, and generalities. Cookie-cutter approaches don't work, either, for the simple reason that the culture of every organization is unique, just as every country is different. With those caveats in mind, I set out in search of practical tips and insights that would be useful and relevant for any organization—the kinds of approaches that would help cultivate the culture that Jenny Ming described at Old Navy, where she built the brand into a retailing powerhouse.

"I was there from the very beginning," said Ming, who is now the CEO of Charlotte Russe, the clothing chain. "But later, I still considered it a start-up. I think when we were a three-billion-dollar company, someone said to me, 'Jenny, we're not a start-up.' And I said to them, 'I think we have to have that mentality of a start-up, because I think it's very healthy to think that way—"Resources are scarce, so what should we do?"' I have learned that you always have to have a little bit of that. It's a state of mind. I think it makes you hungry. It makes you the underdog. You want to prove that you can do it."

If you scan the list of CEOs at the start of this book, you'll notice that a large number of them run technology companies. That is no accident. Successful tech firms tend to grow quickly, giving their leaders a keen sense of the cultural challenges that crop up as they add more employees and layers of management.

Tech CEOs also often apply the same innovative thinking to culture that they use to develop products. For example, Phil Libin, the CEO of the software firm Evernote, once asked his wife for advice on what he could do that would have a big impact on his employees' lives. Her response: free housecleaning twice a month for every staff member. Libin took her advice and also did away with a formal vacation policy, but to offset the pressure some employees might feel to not take a break, he gives each staff member $1,000 spending money to go away on real vacations. (Visits to the in-laws don't count.)

And the leaders of technology firms have to think hard about culture because they are in a war for talent—not just to attract employees but also to hold on to them. After all, any decent software engineer gets e-mails every day from headhunters offering raises and new perks to jump to another firm. So CEOs have to create a place where employees want to stay.

As the old-school approach of command-and-control leadership fades, companies in all industries will inevitably move in the same direction as these tech firms, and try to tap into the deeper passions of employees.

"I think we're at this evolutionary time in business where it's all about people," said Lily Kanter, a former Microsoft executive who cofounded the home furnishings company Serena & Lily. "And we have to embrace that and embrace people's purpose and their souls to be successful in business. Because if they're just coming to work to be a body, and they're on a treadmill all day long, if you don't tap into what is really meaningful for them in life, you won't have a happy culture."

Talk to enough leaders, and you will hear many of them boast that they have no politics at their firms. That's not realistic, of course, because every company, every organization, every team, has politics. A more realistic approach is to recognize that politics is a fact of life, and that the goal should not be to eliminate politics, but instead to encourage the good kind of politics and discourage the bad—and yes, there are parallels with cholesterol. With that goal in mind, each chapter in this book will focus on a big driver of culture, with insights from top executives on how to avoid some of the usual pitfalls, and instead turn culture into a competitive advantage. To provide a framework, I've organized the book into two sections. The first part will identify the essential elements of an effective corporate culture, and the second will explore leadership strategies for building on that foundation and fostering innovation.

I've used the same approach that many readers told me they appreciated in my first book, *The Corner Office: Indispensable and Unexpected Lessons from CEOs on How to Lead and Succeed.* Each chapter is structured much like a dinner party conversation, with me as the host, guiding the conversation with a large group of CEOs. I will introduce the themes, make some broader analytical points to steer the discussion, and then let the CEOs share their insights, lessons, and stories, in their own words.

Of course, given the churn and creative destruction in corporate America, some of these leaders have moved on from the jobs they held when I interviewed them, for predictable reasons: their start-up didn't take off as they expected; they wanted a change; or perhaps their performance fell short of their board's expectations. Does that diminish the value of their insights? I don't believe so. The CEOs I interviewed are not perfect. They have their strengths and weaknesses, and career peaks and valleys, just like all of us. But their observations about culture, learned through the day-to-day work of leading teams, transcend the ups and downs of stock charts, balance sheets, and job changes, and have become even more valuable as pressures to innovate increase.

Readers will find that some insights in this book will resonate more than others. That's to be expected, as every culture is different, with its own particular challenges and strengths. Just as with leadership, we all have to make sense of culture on our own, in a way that feels authentic for our organizations. The leaders I've interviewed have hundreds of years of combined experience and wisdom, and I'm confident that their smart insights will help all readers make their own companies more quick and nimble, so that they can thrive in this relentlessly challenging global economy.

PART ONE

SETTING THE FOUNDATION

1.

WHY CULTURE MATTERS

Culture eats strategy for breakfast.

It is hard to pin down the origin of that often-quoted expression. Peter Drucker, the legendary management thinker, is frequently cited as its author, but the Drucker Institute says it has no record of him writing or uttering those words. Whoever came up with it, the saying is clever shorthand for the notion that while a smart strategy is necessary to succeed, culture is the X factor that will determine who wins in the long run.

This idea is at the core of how Stephen I. Sadove, the CEO of Saks department stores, leads. As he explained, his approach is the opposite of how Wall Street analyzes companies.

"I have a very simple model to run a company," he said. "It starts with leadership at the top, which drives a culture. Culture drives innovation and whatever else you're trying to drive within a company. And that then drives results. When I talk to Wall Street, people really want to know your results, what are your strategies, what are the issues, what it is that you're doing to drive your business. They're focused on the bottom line. Never do you get people asking about the culture, about leadership, about the people in the organization. Yet, it's the reverse, because it's the people, the leadership, the culture, and the ideas that are ultimately driving the numbers and the results."

It's a lesson that Joseph Jimenez, the CEO of the pharmaceutical company Novartis, learned when he was a division president at another company earlier in his career.

"I was sent in to turn the division around after four years of underperformance. It was a declining business. And when I got there, I completely misdiagnosed the problem. I said, 'Look. We're missing our forecast every month. What's wrong?' I brought in a consulting firm, and we looked at what was wrong. And the answer was that we had a bad sales and operations planning process, where salespeople, marketing people, and operations people were supposed to come together and plan out the next eighteen months and then forecast off of that. So I said, 'Okay. We're going to fix this. We're going to have the consulting team come in and help us make that a better, more robust process, with more analytics.'

"And it turned out it wasn't at all about analytics. Because once we did that, and we put that new process in place, we still continued to miss forecasts. So I thought, 'Something's really wrong here.' I brought in a behavioral psychologist, and I said, 'Look, either I'm misdiagnosing the problem or something's fundamentally wrong in this organization. Come and help me figure it out.' She came in with her team and about four weeks later came back and said, 'This isn't about skills or about process. You have a fundamental behavioral issue in the organization. People aren't telling the truth. So at all levels of the organization, they'll come together, and they'll say, "Here's our forecast for the month." And they won't believe it. They know they're not going to hit it when they're saying it.'

"The thing she taught me—and this sounds obvious—is that

behavior is a function of consequence. We had to change the behavior in the organization so that people felt safe to bring bad news. And I looked in the mirror, and I realized I was part of the problem. I didn't want to hear the bad news, either. So I had to change how I behaved, and make these discussions more of a chance to say, 'Hey, thank you for bringing me that news. Because you know what? There are nine months left in the year. Now we have time to do something about it. Let's roll up our sleeves, and let's figure out how we're going to make it.' It was a total shift from where we had been previously. So after that experience, I always ask all of my people, and I always think to myself, 'Are we really fixing the root cause of this problem, if there's any problem? Or are we fixing the symptoms?'"

That is why culture matters. A successful culture is like a greenhouse where people and ideas can flourish—where everybody in the organization, regardless of rank or role, feels encouraged to speak frankly and openly and is rewarded for sharing ideas about new products, more efficient processes, and better ways to serve customers.

Without that kind of culture, without a sense of shared values and some basic rules for working together, people can easily forget they are part of a team and start protecting and pursuing their own parochial interests. Tony Hsieh, the CEO of the online retailer Zappos.com, said that his early experiences in a start-up taught him the dangers of simply letting culture evolve on its own.

"After college, a roommate and I started a company called LinkExchange in 1996," he said. "It grew to about a hundred or so people, and then we ended up selling the company to Microsoft in 1998. From the outside, it looked like it was a great acquisition, two hundred and sixty-five million dollars, but most people don't know the real reason why we ended up selling the company.

"It was because the company culture just went completely

downhill. When it was starting out, when it was just five or ten of us, it was like your typical dot-com. We were all really excited, working around the clock, sleeping under our desks, had no idea what day of the week it was. But we didn't know any better and didn't pay attention to company culture. By the time we got to a hundred people, even though we hired people with the right skill sets and experiences, I just dreaded getting out of bed in the morning and was hitting that snooze button over and over again. I just didn't look forward to going to the office. The passion and excitement were no longer there. That's kind of a weird feeling for me because this was a company I cofounded, and if I was feeling that way, how must the other employees feel? That's actually why we ended up selling the company.

"Financially, it meant I didn't have to work again if I didn't want to. So that was the lens through which I was looking at things. It's basically asking the question, 'What would you want to do if you won the lottery?' For me, I didn't want to be part of a company where I dreaded going into the office. So when I joined Zappos about a year later, I wanted to make sure that I didn't make the same mistake that I had made at LinkExchange, in terms of the company culture going downhill. So we really view culture as our number one priority. We decided that if we get the culture right, most of the stuff, like building a brand around delivering the very best customer service, will just take care of itself."

The critical importance of culture is a lesson that Steve Case, the cofounder of AOL who is the CEO of the investment firm Revolution, learned in the wake of the failed merger of AOL and Time Warner.

"They were both terrific companies. I think everybody saw it as a big idea to bring them together and help Time Warner move into the digital age and help AOL move into the broadband age. Well, it didn't happen. Ultimately it came down to poor execution of what I thought was a good idea, and that was largely attributable to people and relationships and resentments and pride and egos. There were some substantive strategic debates, but it was mostly about people and trust and relationships. I've seen where that focus works well. It can really accelerate and animate a company to do things that nobody thought was possible.

"Ultimately I think the core lesson is that it's all about people, and so you've got to focus on that to understand what's going on, what the context is and make sure you get people aligned around the right priorities. If you do that well, a lot can happen. If you don't do that well, not much can happen."

∾

Most people—except those who are congenitally cynical or deeply jaded—start new jobs with a sense of optimism and energy, eager to make the most of a new opportunity. They want to make a good impression, get to know their colleagues, and quickly get up to speed on their jobs. They want to fit in with the team, speak up at meetings, and figure out how to help their boss and their colleagues. They want to make important contributions that confirm the good judgment of the people who hired them.

For leaders, the challenge is to create a culture that preserves this energy, and to beat back the forces that make people shut down and daydream about working someplace else.

Marjorie Kaplan, the president of the Animal Planet and Science cable networks, talks with her staff about the need to hold on to the rush of energy that people bring to new jobs—their "best self," she calls it.

"I want us to be the place where, when you come to work, you feel like you have the opportunity to bring your best self, and you're also challenged to bring your best self. And I'm explicit about that. We have those conversations regularly. For a team off-site, I sent out a note to people to say that 'We need to be as fresh as we were when we were new, and as brave as we were when we had nothing to lose.' And that was the focus of the day.

"I think part of management is bringing people along. I want people to feel brave about their ideas. It's really about saying, 'Bring your best self.' Bring your best self every day to work. Bring your best self to the conversation. Bring your best self to the presentation. And we will give you something back. We're investing in you. You're investing in us, we're investing in you."

Laurel J. Richie, the president of the WNBA, also subscribes to a "best self" approach to leadership:

"I want people coming in every day thinking this is a place where they can bring their very best, and I believe that if they feel that way, they will actually do it. I just don't believe in terrorizing, intimidating, testing, catching people off guard. I don't play games. Life's too short and we've got too much to do. I want people focusing on the work, not how to navigate politics. It's my job to create an environment where they can bring their best selves, and good things will happen as a result."

Unfortunately, many people experience a very different dynamic. Posturing, power grabs, and perceived slights can dampen their initial enthusiasm until they start checking their best selves at the door when they walk in to work.

Perhaps some of these scenarios sound familiar: A colleague dismisses an out-of-the-box idea at a meeting with a wave of his hand. A manager ignores an e-mail, filled with ideas for tackling new challenges, sent by one of his employees. A top executive never keeps a promise to have a follow-up conversation on an important matter. Bad behavior is tolerated from certain people but not others. Small meetings are held before or after larger, scheduled meetings, leaving participants to wonder about the disconnect between public discussions and private whisperings.

Or these: Underperformers are left in their jobs because nobody has the stomach to move them out, forcing colleagues to work around them. A manager gives candid feedback only once a year in a performance review, during which she dredges up a misunderstanding from eight months earlier that was never discussed in the moment. Because the overall strategy and goals of the organization are unclear, employees focus on protecting their respective turfs.

Or how about these: Colleagues communicate mostly by e-mail, huddling all day long in cubicles and offices behind their monitors, instead of face-to-face. When they gather for meetings, they pay more attention to their smartphones and iPads than to one another. The sense of possibilities that once fired their imaginations is replaced with shoulder shrugs and eye rolls. They start thinking how it must be better somewhere else. People wander in late, leave early when they can, and dream about vacations.

Even a well-meaning manager may be pulled into the quicksand of such cultures, until he's so overwhelmed that he starts rationalizing the limits of his ability to shape the culture. "You can't expect everyone to be friends." "They call it work for a reason." "People are who they are, and they don't change." He starts making assumptions about why certain employees behave the way they do—they're just lazy, or they're angry because they were passed over for a promotion—and he avoids confronting them about their subpar performance. The manager starts wishing there were some way to shake up his staff, to get their pulses racing.

So it's no surprise that "start-up" cultures hold such great appeal. To be sure, Tony Hsieh's early experience with LinkExchange is a good reminder that a start-up's culture can be just as flawed as a large company's. Even so, it's easy to imagine a start-up at its best: The entire team pulls in the same direction. No walls or silos divide departments. Everyone learns quickly as they take on new responsibilities and wear many hats. Politics and bureaucracy are kept to a minimum. There's shared commitment to do whatever it takes to finish the work. Egos are checked at the door, and people are honest with their feedback because achieving the collective goal matters more than individual feelings. The employees are excited about new possibilities and opportunities, and they work hard not because they have to, but because they want to.

So why is it so difficult to hold on to that start-up spark? The simple reason is that growth works against it; a bigger staff requires more processes and organizational structures. And as jobs become more stable and secure, people inevitably get set in their ways. Roles become specialized and narrow, so employees focus on just doing the work assigned to them, rather than learning new skills. Hallway chatter dies down as people settle into their offices and cubicles. Risk aversion replaces risk taking, so that innovation is choked off.

As companies race to become more innovative, to spur growth in this struggling and fiercely competitive economy, CEOs learn that naming a chief innovation officer or holding more brainstorming sessions won't get the job done. Real innovation happens when all employees bring their best selves to work every day and freely share new ideas to help the team, knowing that they will be encouraged and rewarded for doing so.

The formula is simple but hard: innovation is the by-product of an effective culture.

"If you really want to build something that's going to be around for a very long time and be stable and grow, culture has to be paramount," said Robert L. Johnson, chairman of the RLJ Companies, an investment firm.

"People have to know how your culture operates and works. And once they get it, they adopt it, and it becomes second nature to them that certain things are not done in this company. And that, to me, is one of the attributes of really great companies. The culture is almost like a religion. People buy into it and they believe in it. And you can tolerate a little bit of heresy, but not a lot.

"That's why I think it's easier in that kind of culture to introduce new products and new technologies or new services. When the culture breaks down, it's real hard to be innovative because you've got built-up barriers to interpersonal communications. And if you've built up barriers to interpersonal communications, you've certainly got barriers to new ideas, because that old zero-sum-game mentality sets in, where somebody might think, 'Well, if you guys are going to market it this way, my distribution chain is going to lose power. So I've got to protect my part of the company. And, by the way, I don't like you anyway.' So innovation slows down, and changes don't happen as rapidly as they should and you get that rigidity.

"At the end of the day, people make up companies. And if the culture allows for a lot of interaction and a lot of free-flowing ideas that are not considered threats to anybody, that company will be more innovative."

2.

A SIMPLE PLAN

> *I would not give a fig for the simplicity this side
> of complexity, but I would give my life for the
> simplicity on the other side of complexity.*
>
> —Oliver Wendell Holmes Jr.

That elegant tribute to simplicity as achievement—a favorite quotation of David Barger, the CEO of JetBlue—speaks to one of the most important roles of a leader: to boil down an organization's many priorities and strategies into a simple plan, so that employees can remember it, internalize it, and act on it. Barger went through just such an exercise with his directors.

"We got the board together several years ago to discuss best practices and strategy," he recalled. "Out of that, we ended with twenty-three objectives, four pathways. Fast-forward, and the twenty-three became fourteen in Year Two, ten in Year Three, and then we crystallized them into two—culture and offerings. We have to get a little more specific when we talk about culture, of course, like preserving a direct relationship with our crew or building talent into the organization. These used to be stand-alone objectives. But it's just so much easier now to communicate the two and then to get more specific, as necessary. It's been very helpful.

People on the front lines know what you want them to do, and it's easier to set the expectations.

"You have to be able to simplify things that are complex. At the end of the day, if the thirteen thousand people on the front lines don't understand what you're trying to do, forget it. You don't stand a chance of making it work."

There is no formula, of course, for boiling down strategies and priorities. Each company is unique, and every leader has to figure out the handful of goals and metrics that will get everyone pulling in the same direction, with a clear sense of how their work contributes to those goals. As in many aspects of life, it is hard to simplify things. But developing a simple plan can have a huge payoff, and getting it wrong, with people focusing on goals that don't drive the organization forward, can have a similarly outsized impact. As Peter Drucker once said, "What gets measured gets managed."

Mission vs. Plan

Developing a mission statement can have many benefits. To be sure, many mission statements are wordy and feel forced. But done right, these ambitious, call-to-arms goals can make everyone feel like part of a lofty enterprise and inspire them to achieve things they might not have thought possible. Mission statements' value as a leadership tool is captured in a remark attributed to Antoine de Saint-Exupéry, the French writer: "If you want to build a ship, don't drum up the men to gather wood, divide the work and give orders. Instead, teach them to yearn for the vast and endless sea."

For tech companies that compete for engineering talent, mission statements can make a difference to a prospective employee considering competing job offers, said David Sacks, the founder, chairman, and CEO of Yammer, the social network service for business.

"In Silicon Valley right now, probably the thing that helps recruit people more than anything else, beyond the compensation package that they can get from a number of companies, is the mission," he said. "They have to be on board with the mission of your company. To me, a good mission statement is one that you can boil down to a pretty short phrase—in our case, we want to revolutionize the way people work by bringing social networking into the enterprise. And people can kind of absorb that and they can think about the impact that social networking has had in their personal lives, and then think about the impact that it's going to have on the workplace. A lot of people get inspired by that. I think every good company has a very strong mission statement."

For all the benefits of an effective mission statement, though, its usefulness in driving results—and keeping employees focused every day—goes only part of the way. What is also needed, and is even more important, is a simple plan that everybody understands, so they can see a clear link between the work they do and how it drives those goals.

Tracy Streckenbach is a former Ernst & Young consultant and now the CEO of Hillview Consulting. She helps companies develop and implement turnaround plans. Unlike consultants who deliver a report and then move on, she often joins client companies as a top-ranking executive to get them on the right track. She has seen firsthand the value of a simple plan; to her, such a plan matters more than a compelling mission statement.

"Honestly, I think that a mission statement is less important than every employee understanding what the company's positioning is," said Streckenbach. "To me, the mission can be a little academic. I've got to create change quickly and drive results. When I go into a company, I might ask ten senior executives what they think is unique about the company, and I get very different answers, sometimes even conflicting answers. So just getting everyone on the same page about the company's positioning is more important than the mission. Culture is about performance, and making people feel good about how they contribute to the whole.

"You want to create an environment where people want to be at work. I lived through that whole Internet craze where you couldn't hire people fast enough. During those days, you thought of culture as Ping-Pong tables and disco balls. Now I think the big focus on culture, particularly in a down economy, is on how you get people invested so that they care about what they're doing and feel like they have a hand in things. The only way you can do that is if you have very clearly defined and measurable goals. Then you make sure each and every department knows them, and how their work will support the overall goals.

"It sounds easy and simple, but it's not. In one company, it probably took me six months to clearly define the right goals and how to measure them. It's also devastating if you get them wrong, because then you're encouraging the wrong behaviors. But once you get it right, you see this change in people. They want to get the job done, and not just put in the time."

Few vs. Many

What's the right number of measurable goals for an effective and simple plan? There's no hard-and-fast rule, but there is a clear benefit to having three or fewer: people can easily remember them. Anyone who has ever had to run to the grocery store to pick up a few things knows this rule firsthand—if you have to remember more than three items, you probably should reach for a scrap of paper to make a list, or develop a mnemonic to help you recall what you have to buy.

Shivan S. Subramaniam, the CEO of FM Global, a commercial and industrial property insurer, developed a simple operating framework with just three measurements. The company was the result of a merger in 1999, and the new leadership team felt it needed everyone on the same page.

"A big-picture lesson I learned is that you have to make very sure everybody in the company has the same goal in mind," he said. "It matters less what people do or how they do it, but do we all agree on the same goals? Over the years, that has led to us having very simple goals at our company. We call them 'key result areas' or KRAs. We have three KRAs, nothing terribly fancy, and everybody focuses on them. One is profitability. One is retention of existing clients. And one is attracting new clients. That's it.

"You can talk to our employees in San Francisco, Sydney, or Singapore, and they'll know what the three KRAs are. All of our incentive plans are designed around our KRAs, and every one of those KRAs is very transparent. Our employees know how we're doing. And, most importantly, they understand them, whether they're the most senior managers or file clerks, so they know that 'if I do this, it helps this KRA in this manner.'

"We had people who were in fairly small to medium-size companies before the merger, and now they were part of a large company. They wanted to still be treated the way they were treated when they were in a small company. The kind of communication system that we developed helped enormously."

Joseph Jimenez described how he applied an important leadership lesson that he learned during his college years to his role as CEO of Novartis:

"When I was at Stanford, I was a swimmer, and I was a captain in my senior year. The first thing I learned when I was captain is that you have a lot of people on a team who have different agendas, different objectives. We had to get everyone aligned around a common goal, and the one we set for ourselves was to break into the top five at the NCAAs. In my freshman year, we were number twenty in the U.S. By my senior year, we ended up third.

"When I first became CEO of Novartis, I said: 'We have a hundred and twenty thousand people. That's a lot of people to try to align. The first thing I have to do is to have people understand where I'm going to take the company. And it has to be crystal clear. And not only does it have to be crystal clear, but everybody in the organization has to understand it, they have to have line of sight to that goal, and they have to understand how what they're doing is going to help us move into the future.'

"Throughout my career, all my performance reviews had one thing in common, whether the results were good or the results were bad. They all said that I have the ability to look at very complex situations and make them simple. And I personally believe that if you can't hold something in your head, then you're not going to be able to internalize it and act on it. At Novartis, our business is very complicated. But you have to distill the strategy down to its essence for how we're going to win, and what we're really going to go after, so that people can hold it in their heads—so that the guy on the plant floor, who's actually making the medicine, understands the three priorities that we have as a company.

"We are going to win through science-based innovation— that's kind of the overriding theme. So the three priorities are, first, extending our lead in innovation. And we measure that across all divisions by the number of new compounds that we have approved. The second is called accelerating growth, and that's turning that innovation into sales and profit growth. The way that we measure that is, what percent of our portfolio, in any division, is driven by products that have been introduced in the last three years? And for the company, it's twenty-five percent. For a fifty-five-billion-dollar company, that's a huge number. Driving productivity is number three. That's about getting smarter about the way we're spending our money. So we talk about those three things every quarter. How did we extend our lead in innovation? How did we accelerate growth? And how did we drive productivity?"

In-group vs. Out-group

Once the goals for the organization are set, many companies ask employees at all levels to develop and share their personal plans for contributing to those goals. David Sacks of Yammer, for example, developed a system he calls "MORPH." He explains the acronym here, and the thinking behind it:

"MORPH stands for Mission, Objectives, Results, People, and the H is for 'How,' as in, 'How did you do by the end of the quarter?'

"'Mission' is just a one-sentence description of 'What's your mission at the company? What do you have ownership of?' And that really gets people to think about, 'Okay, what is my overall mission here?' 'Objectives' are the top three, maximum five, things that you want to achieve this quarter. 'Results' are about the metrics you're going to use to measure those objectives. 'How do we know if we've achieved them?' 'People' refers to 'What changes do we need to make in the organization to achieve this? Do we need to hire people? Do we need to create new teams? Do we need to change the way that a team is defined?' And then at the end of the quarter we just ask, 'How'd you do?'

"So I do this first. Then I present it to my direct reports. Then they present theirs to their reports. And it's supposed to be pushed all the way down through the organization. So it's an exercise in alignment. And this is all something that fits on one page. I feel like if you can't get this down to one page, then it doesn't work. It's called MORPH because I worked out the letters to be a synonym for change, because I want to convey to people that management isn't about keeping a lid on things. It's about progressing, changing things.

"We're constantly adjusting what the objectives are, what the goals are, maybe the metrics. So by the end of the quarter the MORPH might have changed and the new MORPH hopefully won't be this complete exercise out of the blue. It'll just be a continuation of what we're already doing. But I think it is helpful just to pause once a quarter and just kind of step back for a second and say, 'What are we trying to do here?' You have to be centralized with respect to direction, decentralized with respect to execution."

At the consulting firm Fahrenheit 212, CEO Geoff Vuleta likes to measure work over one-hundred-day stretches. He described the thinking behind his system, and how it works, weaving into his explanation some insightful observations about human nature.

"One of the traits of a good leader is being able to build loyalty beyond reason, and getting people totally believing that something's possible," he said. "And I've always believed—and this is fundamental to leading a group of people—that everybody wants to be led. They want to know two things. They want to know what they should be doing, and they want to know that what they're doing is important. And you must, therefore, set up an environment in which they totally trust that.

"So your consistency of behavior is the most important thing in running a group of people. We get together every one hundred days as a group, and we draw up a list of all the things that we want to get done in the next one hundred days. And you go away as an individual and come back with commitments to how you're going to contribute to that list. Then you sit down with me and our president and we discuss your plan. It's just our job to make sure that the sum of everybody's plan nails the firm's list.

"And the firm's list is made up of really simple things. What were the things that went wrong in the last one hundred days? Let's get rid of those. You want to nail your pain points and go, 'Okay, what needs to be done to make sure that doesn't happen again?' So that's part of it. And what do you want to do about your brand? How are you going to advance 'thought leadership'? Not all projects are born equal—there are some that are grander than others. What are you going to do to invest in those? Who's going to take responsibility for them?

"And then there's all the personal growth stuff, which everybody includes on their list. You want to advance. You want to grow as a person. There are things you want to get better at. But the thing that's material about the list is that the company has agreed that those things are important. It's a bit goofy to do it the first couple of times because people obsess over how they're going to do something or what they're going to do. It isn't about any of those things. It's only about an outcome. It's only about what will have been achieved within the one hundred days or at some point during the one hundred days. The one-hundred-day time frame works brilliantly because you can never be more than one hundred days wrong as a company. You've got to allow time for people to feel the pain of getting something wrong. And when you create a competitive environment that has total transparency like we do, you won't do it twice. You just won't.

"So the one-hundred-day plan meetings start off with you actually reporting on yourself. You stand up with your one-hundred-day plan, and there's no wiggle room on it. They're outcomes. You did them or you didn't do them. You're enormously exposed because you offered to do it, and you're going to do it. Fahrenheit has only had to fire three people, because the one-hundred-day plan sorts it out beforehand. There's nothing that I'm doing that anybody wouldn't understand or appreciate, because everything's exposed to everybody else. Everybody can see what everybody else is doing. If stuff happens that prevents you from being able to do it, you lean in quickly and either take it off your list or replace it with something else. At no point does anybody in the company not know what everybody is doing in the company, what they've committed to, and what the company thinks is important."

Dennis Crowley, the CEO of the social networking site Foursquare, prefers a weekly system of sharing every employee's priorities, including his own. He developed a system based on "snippets" that he learned about while working at Google.

"Every Monday you send in a bullet list of the stuff you've been working on, and the software compiles a list and mails it out to the entire company," he explained. "So you can quickly scan those lists to find out the status of a project or what somebody is working on. It gives you a nice general overview of the company. So you follow the people you want to get updates from, but we make sure that everyone automatically gets them from me and our COO and our head of engineering and our head of product.

"When I send out mine, the first heading is 'Things I'm Psyched About,' and the next is 'Things That I'm Not So Psyched About' or 'Things I'm Stressed About.' The next thing is usually a quote of the week—something I heard from one of our investors or maybe overheard from an employee—and then I have my snippets below that. The system works out great. I get a lot of feedback from employees. It only takes them a minute or two to read, and it's like a bird's-eye view of what I think is going well at the company and areas where I think we could improve. It's also a good way to start a conversation. So I might write, 'Hey, I heard someone say this, and so let me address why we're thinking about it this way.'"

Kris Duggan, the chief strategy officer of Badgeville, which designs game-based programs for companies, developed an approach he calls interlock.

"I think organizations have a hard time communicating up and down the chain of command and getting everybody mobilized to focus on the same goals," he said. "I've experienced that firsthand—whatever your task or scope of work is, you don't know how that connects to your manager and your manager's manager, and how that is all kind of interconnected.

"So the biggest thing we focus on is this concept of 'interlock.' Interlock is about how we get all the departments connected with their goals—all the way from the CEO to the frontline person—so that all of those goals and controls are transparent. Everybody should know what everybody's goals and controls are, and everybody should understand their individual ones relative to their department's, and their department's goals relative to the company's.

"What we've done to achieve that is to actually publish company-wide goals and controls. We have six major goals this year, and there might be three or four metrics for each of the six goals. And we publish that every month for the entire company. And then we talk every month with the company about whether we're at green, yellow, or red on any of those six things, and we are very transparent about that. We don't hide the bad news."

This total transparency has many benefits. Not only does it clarify the goals of the organization, but also it helps employees see how their work, and their colleagues' work, contributes to achieving those goals. That knowledge reduces uncertainty, which can be distracting. It also helps employees see the big picture, so they can act more like owners—and think about what's best for the organization overall—and less like role players.

Ryan Smith, the CEO of Qualtrics, a software company, has made extreme transparency a hallmark of his company's culture.

"We're not a transparent culture so that we can be cool," he said, "and it's not about an open environment, because that's not what makes a company transparent. It's more around the fact that everyone needs to know where we are going and how we are going to get there. So we want everyone to understand our objectives and make that available to everyone as we're evolving, so that people aren't guessing and they're not internally focused, because that's one of the obstacles that a lot of companies fall into.

"We want to be transparent because we want to encourage our people to have all the information to keep them focused on what really matters: our objectives and how they're going to contribute. So we took our best product guy and some of our best engineers and built a system internally to help scale our organization by knowing everyone's objectives in the company. We have five objectives annually for our company, and everyone goes into the system each quarter to put in their objectives that play into those broader goals.

"We have another system that sends everyone an e-mail on Monday that says, 'What are you going to get done this week? And what'd you get done last week that you said you were going to do?' Then that rolls up into one e-mail that the entire organization gets. And so if someone's got a question, they can look at that for an explanation. What I've found is that when everyone's rowing together toward the same objective, it's extremely powerful. We're trying to execute at a very high level, and in order to do that, we need to make sure that everyone knows where we're going."

The benefits of such an approach can also be explained through neuroscience. David Rock, who coined the term "neuroleadership" and is the director of the NeuroLeadership Institute, has drawn effective leadership strategies from brain research. He explained how shared goals that create a sense of teamwork—in contrast to employees simply trying to protect their turf—can fundamentally change how people perceive their colleagues.

"We make a decision about each person we interact with that impacts basic processing and many other things," Rock said. "And the decision we make about everyone is 'Are you in my in-group or in my out-group?' Now if you decide that I'm in your in-group, then you process what I'm saying using the same brain networks as thinking your own thoughts. If you decide I'm in your out-group, you use a totally different brain network. So the very level of unconscious perception has a huge impact based on this decision of, 'Is this person similar to me? Are they on my team? Do we have shared goals, or are they in my out-group?'

"This is the neurobiology of trust in a sense, but also of teamwork and collaboration. If you can create shared goals among people, you can create quite a strong in-group quite quickly. When you can find a shared goal, you turn an out-group to an in-group. Unless a leader creates shared goals across an organization, an organization will be a series of silos. That's the inherent way that we live. We naturally think in small groups."

A clear plan that creates shared goals will get everyone moving in the same direction and foster a sense of teamwork so the company can execute its plan quickly, and then shift direction when the need arises. Effective leaders recognize that their job is to provide employees with a simple answer to a simple question: "Where are we going and how are we going to get there?" They also know that getting this right is harder than it looks.

3.

RULES OF THE ROAD

> *When you are clear about values, it empowers*
> *people and creates an entrepreneurial spirit that*
> *gives people space and rope, rather than being in*
> *a command–and–control environment.*
>
> —Mark B. Templeton,
> CEO of Citrix

Citrix is a technology firm, so it's hardly surprising that Mark Templeton reaches for a computer metaphor to explain his views about culture:

"The way we define culture overall is 'how companies get things done.' If you have a factory, you get a lot of things done through machinery. Most companies in software get things done through people. So our machinery is people, and to put it in technology terms, people are the hardware and our values are the operating system.

"So the culture starts with people with a common operating system around values and then, once you have that, you can build processes around how you actually get things done on top of that. But clarity around the hardware and the operating system to me is first and foremost."

Citrix's culture is based on three values, Templeton says: respect, integrity, and humility.

"I think people generally want to belong to something of greater purpose that's larger than they are. They're just waiting for it to come along. And I think a culture around values is part of that. People say, 'I want to be on that team, that club, because they believe in something and I actually believe in that, so I want to belong to that.'

"Everyone feels that they're birds of a feather because of common values. We have clarity around where we're going, and then they get to fill in how we're going to get there—with the right kind of management, of course, and leadership, and the right kind of processes and metrics. But it's very much seen as a giant start-up."

Templeton makes a compelling case for the importance of codifying values. When employees are given guidelines for behavior—rules of the road—they can focus more on the work at hand, rather than on navigating the stressful politics that naturally occur when all sorts of bad behaviors are tolerated.

"I think there are two kinds of cultures and then you can subdivide them after that," said Mike Sheehan, the CEO of the Hill Holliday advertising agency. "One is based on a foundation of insecurity, fear, and chaos, and one is based on a firm platform where people come to work and they're worried about the work. They're not worried about things that surround the work and are not important. If leadership doesn't provide a forum for that kind of stuff, it dies quickly. People forget about it and they just focus on doing their job."

Developing values for an organization can seem like a risky proposition; it's easy to imagine how the exercise can seem straight out of an episode of *The Office*. But many leaders see values as an essential component of business—for attracting and retaining talent and for helping people focus on the work.

Two overarching rules about culture and values have emerged from my interviews with hundreds of CEOs:

No. 1: There is no "right" way to develop values. Values can come from the leaders themselves or they can be developed with input from everyone in the company (or a mix of both). What truly matters is . . .

No. 2: The company has to live by its expressed values, reinforce them every day, and not tolerate behavior that is at odds with them.

Here's why: if employees start seeing a disconnect between the stated values and how people are allowed to behave, then the entire exercise of developing explicit values will damage the organization, because it will make people shut down, roll their eyes, and wonder why on earth they hoped that this time might be different.

"I think it's easy for people at many companies to become cynical, which then leads to politics, which can create a cancer that can topple even the greatest companies," said Kathy Savitt, the former CEO of the e-commerce site Lockerz. "And I do think cynicism is that first cell, so to speak, that can metastasize within an organization when you feel a company is not actually living out its core values. A good example is when a team member has a great idea or has a big issue with a customer experience and no one responds, no one even acknowledges it, no one gets back to them. The idea festers, problems continue to mount, no one listens. How does that person not become cynical? That's a recipe for cynicism."

Values can create tough decisions for managers, especially when the behavior of a superstar performer undermines the stated values. But many leaders said there has to be a zero-tolerance policy for people who clearly and consistently flout the rules of the road.

"You have to set a belief system in your organization," said

Steve Stoute, the CEO of the ad agency Translation LLC and the chairman of the beauty products company Carol's Daughter. "Once you do that, if you have people who have not bought into the philosophy, you need to identify them and move them out quickly. It's to their benefit and your benefit. If you ask most executives, they know within the first thirty or sixty days if a person is going to work out, but it takes them seven months to a year to get them out of the organization. That's a waste of time.

"I think that it's very important, no matter how big you get, to have checks and balances to know when somebody has not bought into the culture, because at some point in your organization, something is going to backfire and something's not going to get done because somebody's not paying attention. The beliefs of the organization are not going to be passed along because you have people who have not even bought into the belief system. And here's the biggest problem: bad behavior is contagious. And once that starts hitting a company, no matter how big you are, no matter how small you are, that will start the demise of a great organization.

"Bad behavior is the blatant act of ignoring the belief system of the company—it means not paying attention to the strategic intent of the company and not being aligned around the goals of the organization. So when somebody has not bought into the system, that becomes very contagious, it becomes a cancer in the organization, especially when you're talking about mid-level talent. Because any organization is not going to move forward unless mid-tier management helps foster young talent to become better. And if that doesn't happen then you are actually going to lose talent."

Robert Johnson of the RLJ Companies is also adamant about having little tolerance for behavior that doesn't conform to the stated values of an organization.

"Culture is like a circle, and great companies won't even tolerate a superstar going outside the circle," he said. "I can have the greatest sales guy in the world, or the greatest marketing manager. If they go outside the circle, they've got to leave because it is a direct threat to the cultural confidence you're trying to build. And you can't carry out a vision or go on a crusade without the total confidence of everybody who's going on that mission with you."

Crafting Values

As each leader describes the values of his or her organization—and the process for coming up with them—it is important to notice the many different ways companies develop and then reinforce the values.

Here's Steve Stoute describing his approach at Translation LLC and Carol's Daughter:

"Doing what you say is a core value. You can't have people inside a company who are saying things but have no intention of doing what they say. They might have good intentions when it leaves their mouths, but that's exactly where it ends. You have to find those people immediately, because those people hurt a growing organization. So I have people aligned around the idea that we are not going to just say things that we don't mean; we're actually going to do what we say we're going to do.

"Another core value is that you have to have a major and a minor. I don't want you just using your academic background inside the workforce. If you're a photographer, if you are a DJ, if you are a blogger, I want those skill sets as part of what you do inside of our company. As the head of a company that focuses on culture and understanding culture, I need everybody to have a major and a minor inside of my organization. They're embraced, and they're part of the creative process.

"So at Carol's Daughter, somebody could be in customer service, but their minor may be in photography. I want to hear their point of view as we start looking at packaging, as we start looking at how we are going to market to our consumer. I think that's very important. If you have people in your organization who touch culture, you have to reward them for bringing that into the workplace. Most companies mute their employees' 'off the court' activities. I want what you do off the court to be a part of our growing organization. It builds camaraderie inside of a company.

"And ideas can come from anywhere. There are no titles around an idea. As the CEO, I'm the chief editor of the company, but I want the idea to come from anybody. There's no bureaucracy around an idea. In fact, bureaucracy around an idea is the death of an organization. I tell people all the time: If you have a great idea, and you're passionate about it, and it makes sense, and you can't get your boss to hear your idea, then you should leave. That's not an organization that you're going to thrive in. You know why so many companies let great talent out the door? Because there was no platform for great talent to be heard, so they get frustrated and leave."

Lars Björk, the CEO of the data software firm QlikTech, explains his company's five values, and how the company rewards people who best model the behavior:

"The first one is 'Challenge,' because we are a disruptive software company. Always challenge the conventional, because if you follow others, you can at best be number two. And if you want to win, you've got to find your own way to the top. And we challenge each other at QlikTech, because if you're complacent, you're not going to survive.

"The second one is 'Move fast,' because we are building a hypergrowth company. It's okay to make mistakes, just don't make the same mistakes. Learn from them. The third one is 'Be open and straightforward.' What that means is just be open if you think something is wrong. We hear everyone out. It's important that everything is on the table, because somebody might have something brilliant to say. But when we leave the room and we've decided on one thing and your view might not be incorporated in that, you still have to respect the decision.

"The fourth one is 'Teamwork for results.' This is not about the individual. This is about the team, the power of the team. You reach out and you speak to people everywhere, and you learn a lot from people that way, because there are a lot more similarities between cultures than you might think.

"The fifth one is 'Take responsibility.' You're given authority to be part of a lot more than just your position, but some responsibility also comes with it. And if you want to grow fast, you have to put into people's DNA the idea of being cost-conscious. That's why we all still travel coach.

"At our annual company summit each year, we give out awards for each of these five categories. The employees nominate people in each category and then you become a value ambassador for the coming year. It could be an individual, or it could be a team. If we don't live and breathe it, it fails."

Robert LoCascio, the CEO of LivePerson, a software company, describes why he felt the need to codify values in his company, and how he had to let go of the process to develop them. It's also worth noting that the company was able to boil down its values to just two, and that many employees left the company after the values were established.

"The last two years have been a real change," LoCascio said. "We're at about a hundred and sixty million in revenues, but when we hit about a hundred million in revenues and five hundred employees, I could see things evolving in the company that could really hurt us in the long term. We started to become bureaucratic a little bit. You start to add middle-management layers. So this deadweight layer comes in, and their job is just to make sure everyone just does their stuff. And then what happens is you become very siloed in many ways, and the managers want control. You get infighting, and people are more focused on getting offices and titles because they ask, 'It's very hierarchical here, so how do I get up the chain?'

"I remember one moment in particular. We had the cubicles in the middle and offices on the outside, and I walked past an office, and noticed two people shoved inside a small office. And I asked them, 'Why are you guys shoving yourselves in an office? You were just in cubes out here.' And they said, 'Well, we're directors now, and directors get offices.'

"I never believed in culture and core values up to that point. But I really became very reflective because I wasn't so happy with what was happening with the company. I'd seen things that had made me realize we were becoming very traditional. I started to spend more time with a couple other companies, and it made me realize, as a founder, that if I left the company, it could be a totally different place because of the next set of leaders. And that's what kills companies. So I thought that values were the way to have a long-term sustainable company. I finally said, 'It's time to start making the change in the company.' I came in one day and asked all the leaders to move out of their offices, and that was the start of a painful process of change.

"I invited everyone into the process, and I provided context to the whole thing. I remember that I had an all-company call and said, 'You know what? We've done amazing things. We went from twenty million dollars to a hundred million in five years. We should be very proud, and we should congratulate ourselves. And now we are going to move forward, and we need to design a different environment, a different company. So let's acknowledge that we're going to do that as a team, as a company.'

"And we ended up all going to Israel, where our research offices are. All the employees, more than three hundred of them, came. I remember some people said, 'I don't want to come. This is dumb.' There was a lot of friction. But everyone came, and we spent three days doing this cultural evaluation. We were in small groups of twenty, and we sat in circles. The first day people were like, 'I don't know what's going on here. This is kind of crazy. It's a little Kumbaya-ish.' We had forty core values at that point—innovation, customer first, all the typical ones and then a bunch of other ones. Then on the second day we started to get more reflective about what all this meant. We eventually got to two core values: 'Be an owner' and 'Help others.' 'Be an owner' is about us being owners as individuals, driving the business, and 'Help others' is about being reflective and understanding that we're in a community here. We can't be selfish. And so that's where we ended up with our core values, and it was a really fascinating process.

"Over a year and a half, about a hundred and twenty employees left, and I ended up replacing about three-quarters of the management team in the end. Half selected out, and half were let go. What was interesting is when we would be in meetings, somebody would say, 'You're not allowing me to be an owner.' Or I had an employee who came to me and said, 'I'm leaving the company because I don't want this. I can't handle being an owner. I just want to be told what to do.'"

While LoCascio involved his staff in the process, Kip Tindell, the founder and CEO of The Container Store, developed the values himself in the early stages of his company.

"I studied a lot of philosophy at Jesuit High School in Dallas," he said. "One of the things that really struck me was that most people seem to think that there's a separate code of conduct in business from your personal life. And I always believed they should be the same. So we have what we call foundation principles. They are talked about and emphasized around here constantly. They're all almost corny, a little bit Golden Rule–ish, but it causes two things. It causes everybody to act as a unit. Even though we're sort of liberating everybody to choose the means to the ends, we all agree on the ends, and the foundation principles are what cause us to agree on the ends. As a result, we have people unshackled to choose any means to those ends, but it's not mayhem because our foundation principles kind of tie us together. When you start a business you have a very fortunate thing in that you have the opportunity to sort of mold a business around your philosophy.

"The way we create a place where people do want to come to work is primarily through two key points. One of our foundation principles is that leadership and communication are the same thing. Communication is leadership. So we believe in just relentlessly trying to communicate everything to every single employee at all times, and we're very open. We share everything. We believe in complete transparency. There's never a reason, we believe, to keep the information from an employee, except for individual salaries.

"One of the other foundation principles is that one great person could easily be as productive as three good people. One great is equal to three good. If you really believe that, a lot of things happen. We try to pay fifty to a hundred percent above industry average. That's good for the employee, and that's good for the customer, but it's good for the company, too, because you get three times the productivity at only two times the labor cost.

"Another has to do with intuition, where we just beg and plead and try to get employees to believe that intuition does have a place in the workforce. After all, intuition is only the sum total of your life experience. So why would you want to leave it at home when you come to work in the morning? Maybe the most important one is: Fill the other guy's basket to the brim. Making money then becomes an easy proposition. That's sort of the opposite of a zero-sum game, and it means creating a mutually beneficial relationship with everyone we work with."

Although it is obviously much easier to start with a clean slate, new leaders can also introduce explicit values in companies that have been around for years. Cathy Choi, the president of Bulbrite, a lighting maker and supply company, decided to formalize the values of her family's business, which her parents had run for decades.

"One of the first things I did when I became president was to build an intentional culture. We were small, and the allegiance was to my dad. But when we were transitioning over, I made a concerted effort to make the company the leader, not me or my dad. We brought in an outside consultant to do a workshop for the whole company to talk about our value system.

"It started with the idea that values really are personal. We asked the question, 'What is it that you love to do when you're not working?' If somebody said cooking, then we asked why. Maybe they liked the creativity of it, or the excellent result. We put all the different values on the whiteboard and started to see patterns.

"Integrity was important to a lot of people. Team spirit was, too—they really like being committed to each other and investing in other people. So we got ten words or values out of that exercise, and then we tried to whittle them down. We narrowed it down to an acronym: BE BRITE. And each letter stands for the value that's important to us. And when we hire people, we look for people who are aligned with that value system. People who didn't align with us ended up not being able to stay.

"The 'B' is for Bulbrite. The 'E' is for excellence in everything we do. The second 'B' is a better way of doing things—or 'Be innovative.' The 'R' is for relationship building, and the 'I' is for integrity. The 'T' is for team spirit. And the 'E' is about educating yourself and others. I look for people who want to learn and grow. Then we came up with a list of accepted behaviors that support the value system, written by each team member.

"Once a month, we recognize a person for being the most value-driven. They're nominated by the employees. At the end of the month, we look at all the nominations. And at the end of the year, out of all the twelve people who've won throughout the year, we vote on the recipient of the year. Everybody writes something about why they think the person deserves it. And at the holiday party, the winner has to stand there and listen to all the things that were said about them by their colleagues. You can't force people to say things about other people. That just comes from them living the values."

Victoria Ransom, the CEO of Wildfire, a social media marketing software company, said it took a couple of attempts to decide on the company's values.

"We tried really early on in the company, when we had about twenty employees, to codify our values, but we didn't get that far because it felt forced," she said. "But then as we got bigger, we were expecting a lot from our people—that they could just come in and through osmosis figure out what our values were. And wouldn't it be better if we just told them? The values are here already, but let's make it clear what they are, particularly because you want the new people who are also hiring to really know the values.

"Another reason was that we had to fire a few people because they didn't live up to the values. If we're going to be doing that, it's really important to be clear about what the values are. I think some of the biggest ways we showed that we lived up to our values was when we made tough decisions about people, especially when it was a high performer who somehow really violated our values and we took action. I think it made employees feel like, yeah, this company actually puts its money where its mouth is. We also wanted to put in more of a formal procedure for reviews, and if we're going to review people, it's got to be clear what we're reviewing them on when we consider whether they are living up to the company culture.

"My cofounder, Alain Chuard, and I spent a weekend writing down what we value in our people at Wildfire. And then I literally sat down with every single person in the company in small groups and got their feedback—what do you like, what don't you like, let's tweak this. Some companies' values are really about what the company stands for. We took more of the approach of what do we look for in our people. Passion was a very important one. Team player. Humility and integrity. We had courage, and that was all about speaking up—if you have a great idea, tell us, and if you disagree with people in the room, speak up. Curiosity was one of them, too. We really encourage people to constantly question, to really stay on top of what's happening in our industry, to learn what other people in the company are doing. And the hope was to break down these walls of 'them versus us.' Impact—that was really trying to get at a value we have in the company of wanting to measure whether you're having an impact. And the final value is more outward looking, but it was 'Do good, and do right by each other.'

"In the end, it was a much, much harder process than I'd imagined. When I did these sessions with people to get their feedback on the values, most people were really excited. The ones who weren't inevitably came from large companies who had gone through that process before, and they were very skeptical. I think the best way to undermine a company's values is to put people in leadership positions who are not adhering to the values. Then it completely starts to fall flat until you take action and move those people out, and then everyone else has faith in the values and culture again. It can be restored so quickly. You just see that people are happier."

At the cloud-computing company Appirio, CEO Chris Barbin regularly asks his employees about one of the firm's key values to underscore their importance.

"We have three values that we hire against and three that we run it against," he said. "The three that we hire against are trust, professionalism, and gray matter—as in, 'How smart are you?' The three we run it against are customers, team, and fun. That last one is really core—if you're not having fun eight out of ten days on a consistent basis, you've got to say something. You can't just expect that your manager always knows if you're not having fun.

"I reach out to a lot of employees. It's one of the first questions I ask: 'Are you having fun?' I can see it in their eyes, hear it in their voice. I'll just ask, 'What's your ratio of fun days right now? Are you six, eight, nine, are you four out of ten? If you're four, why?' It helps me get to root causes, since it's a pretty easy thing for people to think about."

Symbols and stories are also effective ways to reinforce values. Andrew M. Thompson, the CEO of the biomedical firm Proteus Digital Health, uses both.

"Culture in our company is a really big deal," he said, "and we have a values system built around quality, teamwork, and leadership. One of the activities around that cultural framework is the idea that employees can recognize each other—groups or teams can recognize or be recognized by other employees for doing things that specifically demonstrate those values.

"The way that works is that employees will write a nomination for other employees, and if it's accepted, which it generally is, then at a company meeting the employees making the nomination will stand up and tell a story about how someone or a team of their colleagues was fantastic and here's what they did that was of really high quality, or an example of great teamwork or really strong leadership.

"The people who are recognized get a quarter-ounce gold coin, with the idea being that you can keep it as a trophy or sell it and have a few extra dollars. And the people who make the nomination get to go out to dinner. What I like about this is that management doesn't do this. People do this for each other. It really promotes what I'm going to call mutuality. People spend a lot of time in organizations being focused on hierarchy. The best, strongest, and most functional organizations are ones where the horizontal relationships are really powerful and where people trust each other, work with each other, support each other, help each other, hold each other's hands, and move forward together. You have to build a very high level of trust, and a very mutually respectful organization where people work with each other and where employees are recognizing each other, rather than management doing it."

Russell Goldsmith, a former movie industry executive who is the CEO of City National Bank in Los Angeles, introduced an award program, inspired by *American Idol*, that uses stories to reinforce the bank's values.

"We talk a lot about stories. They're a really important part of how we teach and reinforce the culture, and how we reward behavior," Goldsmith observes. "Maybe it's because I came out of the entertainment industry. If you had talked to me about a project when I was at Republic Pictures, I would have said it's about story. With movies, if you don't have a great script, forget it. One of the things I noticed at City National is that we have a lot of great stories to tell. If you look up City National, one of the stories you will see is the story of Frank Sinatra's son who was kidnapped. The first CEO, Al Hart, was a real friend of Frank Sinatra's and famously opened the vault on a Saturday and got the ransom money. That happened in the early sixties, but people are still telling that story. It's a source of pride.

"We brought in consultants to teach people how to share stories in a more organized way that underscored the culture. We do something called 'Story Idol,' and every quarter there's a competition among our seventy-nine offices. It's a way to give colleagues a pat on the back and a moment in the sun for doing the right thing, and it democratizes and decentralizes positive reinforcement. We then have a Story Idol competition for the year in a big meeting with the top three hundred people in the company. People tell stories about what they did that promoted teamwork or helped a client by going the extra mile. It's like telling stories around a campfire, but they're doing it around conference tables.

"Story Idol is all online, on our intranet. People make submissions—about fifty to a hundred each quarter. It's kind of crowd-sourced, and people vote on the best one. The people who submit the winning stories all get iPads. The winners themselves—the colleagues who are singled out for going the extra mile to help our clients—they get significant cash awards. But what matters most is the recognition, and the respect from your peers as you stand on the stage in front of three hundred people."

To reinforce values, Mark Templeton encourages employees at Citrix to use the values to screen candidates during the hiring process and to call out colleagues on behavior that is at odds with them.

"Communicate the values," he said, "make them simple and clear, and then whenever you observe a violation, no matter what it is, you shine a light on it. Ask everyone to do that if they see a violation, and hold everyone accountable to it, including your manager and your colleagues. And what I also point out to people is that everyone participates in this because they are all on a hiring team. So you have to understand the values because people will come through the filter with a great résumé and all the skills, but have they demonstrated the right cultural characteristics?"

A number of leaders have introduced clever expressions that are picked up and repeated by employees. Among its many other benefits, such shorthand continually reinforces a key value of the company.

At LinkedIn, for example, "next play" has become the unofficial company mantra, said Jeff Weiner, the CEO. "The person I borrowed it from is Coach K [Mike Krzyzewski] of the Duke Blue Devils," he recalled. "Every time the basketball team goes up and down the court and they complete a sequence, offense or defense, Coach K yells out the exact same thing, every time. He yells out 'Next play!' because he doesn't want the team lingering too long on what just took place. He doesn't want them celebrating that incredible alley-oop dunk, and he doesn't want them lamenting the fact that the opposing team just stole the ball and had a fast break that led to an easy layup. You can take a moment to reflect on what just happened, and you probably should, but you shouldn't linger too long on it, and then move on to the next play."

Ben Lerer, the CEO of Thrillist Media Group, which operates men's lifestyle and shopping sites, said that a phrase that is repeated often at his company is "Don't hope."

"What that means is don't wait for somebody to do something for you," he said. "Don't do something ninety percent well and hope that it'll slide through. Don't rely on luck. You have to make your own luck. The only thing you can do is try your absolute best to do the right thing. And then if it doesn't work out, you know there's nothing else you can do. The only time when you can have real regret is when you didn't do everything you could do. I want to never hope, even though I hope just like everybody else. It's just important to know that you're giving as close as you can to a hundred percent, dedicated effort, and you're being thoughtful about it.

"The expression probably came up about five years ago when someone was asking me, 'What's the best piece of advice you can give an entrepreneur?' The first place that I used it was really early on in the business, when there's no way to point fingers, when you're just four or five people, and you have to will everything yourself."

At Medline Industries, a healthcare supply company, employees often hear the president, Andy Mills, tell people that they "have to kiss a lot of frogs."

"It means that sometimes we think something is going to be a dead end, but you just never know," he said. "Maybe the frog is going to turn out to be the prince. So we might be in a meeting and somebody will say, 'What do you think about this?' And my answer will be 'You've got to kiss a lot of frogs.' If I didn't use that expression, I'd probably say that it's a long shot, but who knows? I don't want people to feel bad for trying something that doesn't work out. So that's my way of saying not everything is going to work out."

Chris Cunningham, the CEO of Appssavvy, a social media marketing firm, started using the word "crush" in the offices, and it was quickly adopted by others.

"It's just a word that I've used—'We're going to crush it. We're going to crush it this year,'" he said. "There aren't that many words that I think sort of embody the sense of confidence, that we're going to go for it. And people want to hear that. I think part of leadership is saying, 'I'm going to go into battle with you.' For whatever reason, 'crush' just feels like the essence of what we're going to do. We're going to crush the competition. The next thing you know, you hear another person saying 'crush.' And then they'll sign off 'crush' on e-mail. And then you go into a meeting and three people will be talking about crushing it. It's part of our culture right now. It's funny how this one word has just carried through everything we do. People want that level of energy."

Marcus Ryu, the CEO of Guidewire, which develops software for the insurance industry, said he tries to establish a culture of embracing adversity.

"It's been very hard to get here, and we should take pleasure in how hard it has been. It's hard to build our products. Everything is hard," he said. "So my expression for it is 'tunneling through granite.' I say that we have tunneled through granite to get to this point, and there's an infinite amount of granite left. We'll never get through it all. So you have to decide, do you actually enjoy tunneling or do you not want to be part of this, because I've got nothing to promise you besides an infinite amount of more granite. People have said, you know, it's a little gloomy. It's like you're just saying it's nothing but blood, sweat, and tears forever, for all of eternity. And so I've tried to lighten up about that."

As all these examples show, there is no "right" set of values for an organization, nothing that stands out as a best practice. Values have to emerge in a way that doesn't feel forced, so that they reflect the personalities and beliefs of the leadership team and the collective culture of the organization. They shouldn't be imposed on a company, because they are likely to be rejected like a failed organ transplant. They have to be used as grounds for firing people who clearly violate them. And leaders have to make sure their own actions reflect the values, or they risk creating a sense of deep cynicism among their workers.

Values are perhaps the most explicit manifestation of culture, a code of behavior that everyone can agree on. Without them, companies will drift, said Charlotte Beers, the former CEO of Ogilvy & Mather Worldwide.

"Culture is about how we do things around here," she said. "I have found that most of the time the culture is informal and interpreted by instinct. The most dangerous thing is a company with no culture, where one guy speaks of it one way and one woman speaks of it another. Then you've got a company that's rudderless."

4.

A LITTLE RESPECT

It's really hard on organizations when people in power throw their weight around. It creates an unsafe environment for collaboration. It breeds defensiveness.

—Bob Brennan,
former CEO of Iron Mountain

Unless you are one of a fortunate few, you've had at least one bad boss over the years, someone who was overly and unfairly critical of your work, and maybe even humiliated you in front of your colleagues. We can all remember those moments as if they were yesterday. Many CEOs worked for a lousy boss when they were younger, and the experience deeply informed their leadership philosophy. They want to create a culture of respect, because they know that people will shut down on the job and simply go through the motions if there is a culture of fear in the workplace.

Richard R. Buery Jr., the CEO of the Children's Aid Society, said a bad experience with a former supervisor provided extra motivation to be in a leadership position so that he could influence the culture.

"One leadership lesson I learned—and this goes in the bucket of 'obvious'—is the critical importance of being a good person and treating people well," he said. "I remember one of my two brief jobs as a lawyer, and having a boss who upbraided me in front of a group of my colleagues. And I think substantively she was wrong, but that wasn't the point. As a colleague, you should have enough respect to come and talk to me, even as a boss, to say, 'Look, I want to pull you aside. This is what I think you should have done. This is what you didn't do.' And that always just stayed with me.

"One of the reasons I wanted to be an entrepreneur, frankly, is that I didn't want to have any more bad bosses. I wanted to create an environment where it wouldn't be okay to treat people like that. This is true in every sector, but I think it's particularly true in my sector, of nonprofits. People work really hard for not a lot of money, not a lot of glory. You should be honored for the work that you do, and that's something that I feel very, very strongly about. I love that I get to sort of play a role in making sure that people who treat each other well are the people who are valued."

Shellye Archambeau, the CEO of MetricStream, a firm that helps companies meet compliance standards, said that early experiences taught her the importance of a culture of respect so that people could do their best work.

"I've definitely been in places with cultures that were harsh," she said. "I was at a company once where you went through a whole series of interviews to get hired. Then, once you were hired, you still had to prove something before people would accept you. So what happens in those kinds of environments is that there's more fear than there should be.

"I don't think people perform well when there's fear. Maybe people aren't physically afraid, but they feel fear. And when people are afraid, the whole chemistry in their body changes. You just can't be as successful in that kind of environment. I think the best environments enable people to actually perform their best, but you're still clear about what's expected."

Ben Lerer of the Thrillist Media Group also had a bad experience with an early boss that directly shaped his thoughts on culture.

"I felt very mistreated by a manager I had in a previous job," he recalled. "Part of the problem was that I was young and immature and I sort of walked in on Day One out of college and had this attitude of 'Give me the keys.' But I ultimately didn't like going to work because of the way I was treated; my work suffered, and I didn't have confidence in what I was doing. And ultimately that led me to decide to leave.

"I remember being regularly publicly humiliated. I'd send out an Excel spreadsheet that didn't have first and last names broken out into separate fields, and he sent a 'Reply All' to the entire company telling me how stupid I am and how bad I am at Excel. There were so many situations where I remember being just made to feel inferior and stupid, no matter how hard I worked. I was a kid out of college and I was not qualified to do some of the work I was being asked to do, but I did my best. And when my best wasn't good enough, I was told I was very stupid, essentially.

"And I just remember saying: 'I never want to make anybody feel this way. This is horrible.' And it made me think: I don't want to be in a situation like this again. I want to create a better situation for myself. So we really try to do that. And I think we succeed more than we fail on a person-by-person basis."

David Rock of the NeuroLeadership Institute explains why people are so affected by fear and humiliation in the workplace, and why so many CEOs refer to the "scar tissue" they developed from those experiences.

"There's been a ton of research in the last ten years or so that shows that things that create the strongest threats and rewards are social," he said. "And social threats and rewards activate what's called the brain's primary threat and reward center, which is actually the pain and pleasure center. And this was a big surprise, to see that someone feeling left out of an activity, for example, would activate the same regions as if they put their hand on a hot plate.

"So it's not just a metaphor that these social feelings are sort of like pain. They use the same network in the brain as pain. If your boss disses you in front of a team, every time you remember that for the rest of your life you feel the pain again. That's scar tissue."

The challenge—and opportunity—for leaders is to create a culture where employees understand that treating one another with respect is a bedrock value.

Setting the Tone

Robin Domeniconi, the former chief brand officer for the Elle Group, said she uses the expression "MRI" as a cultural cornerstone.

"I think most of my success with my teams has been built around this idea that none of us owns anything," she said. "We all are here together. So whatever you do, we are all here, as a group, to work together. You have to be able to finish each other's sentences. You have to be in this together. We want the same outcome.

"One lesson I learned is from a phrase I picked up called MRI. It means the 'most respectful interpretation' of what someone's saying to you. I don't need everyone to be best friends, but I need to have a team with MRI. So you can say anything to anyone, as long as you say it the right way. Maybe you need to preface it with: 'I'm just curious, and I want to understand what you're saying better. Right now, my point of view is quite different. So can you help me understand why you don't want to do this, or why you wanted to do this?'

"If you get people talking and challenging each other, you're going to have the ability to arrive at the right decision so much quicker and so much easier. I just make it so it's a human environment. I'm not going to motivate by fear, but I'm going to motivate by saying: 'Let's win. This is going to be so much fun to figure out. Let's figure it out together.' I guess my management style is very much about imagining we're all children and really vulnerable. Because we are. We're all vulnerable, and we all are really human. We have all this stuff inside of us that we've carried with us. So if you have compassion for that, and you understand that, and you know someone's smart, then you need to make an effort to understand why they may behave the way they behave. You need to trust everyone you work with—and it goes into personal relationships, too—because the only thing that creates jealousy, the only thing that creates fear, is that you're not trusting or understanding something. And if you can communicate what your fears are, your challenges are, and if you trust that the people you work with all want the right outcome, then the environment is going to create itself. It really does."

Mark Templeton of Citrix often repeats a saying at his software firm to ensure that people feel their opinions are valued just as much as the next person's, regardless of the title on their business card.

"You have to make sure you never confuse the hierarchy that you need for managing complexity with the respect that people deserve. Because that's where a lot of organizations go off track—confusing respect and hierarchy—and thinking that low on hierarchy means low respect; high on the hierarchy means high respect. So hierarchy is a necessary evil of managing complexity, but it in no way has anything to do with respect that is owed an individual," he said.

"If you say that to everyone over and over and over, it allows people in the company to send me an e-mail no matter what their title might be or to come up to me at any time and point out something—a great idea or a great problem or to seek advice or whatever."

At the mobile technology company 3Cinteractive, CEO John Duffy has made respect one of the core values of his company. He said the zero-tolerance policy on disrespectful behavior frees up colleagues to police one another.

"We have absolutely clear discussions with everyone about how respect is the thing that cannot be messed with in our culture. We will not allow a cancer. When we have problems with somebody gossiping, or someone being disrespectful to a superior or a subordinate, or a peer, it is swarmed on and dealt with. We don't always throw that person out, though there are times when you have to do that. But we make everyone understand that the reason the culture works is that we have that respect. And there is a comfort level and a feeling of safety inside our business," he said.

"We recently did an employee survey that was really intense. It wasn't just 'Are you happy?' It was eleven questions about your happiness, answered on a scale of one to seven. The question that kept me up for a week was 'Do you trust John Duffy?' Not the company, not the mission. I was asking them about me. How do you feel about me? I got more than ninety percent extraordinarily positive responses. So that's where it starts. I have to set the example for treating all of our employees properly, respectfully, and appropriately.

"When they're awesome, I tell them they're awesome. When they mess up, they hear about it. But do it the right way. Do it consistently. Do it with respect. No yelling and screaming, but 'Here's our expectation, and here's where you missed. What do you think you need to do to get better so this doesn't happen again?' That's what creates the positive culture. That's what attracts amazingly talented people."

Bob Brennan, the former CEO of the data services company

Iron Mountain, offered a provocative argument that the traditional hierarchical approach to running a company creates an "unsafe" environment for workers—hardly the environment you want if your goal is to get people to take risks.

"I think businesses are going through this transformation where command-and-control leadership is dead," Brennan said. "The problem is, a lot of managers haven't been told this. They are very much in a command-and-control reflex. That's what they've learned. That's what they've seen in movies. That's what they've seen in their careers.

"So it's important for us to establish a framework that says, 'Here's how we want you to behave.' For instance, our managers need to seek constructive feedback on their performance from the people who report to them. We're not talking about 360-reviews once a year. It should be a constant dialogue in one-on-ones about 'How can I improve my game?' If I'm not seeking that feedback, I'm creating an unsafe environment for you. And it is, in fact, unsafe, if I'm reviewing your work, but I'm not asking you to collaborate with me and review mine. That would presume that I'm fine with the way I'm performing, yet I've been sitting here offering constructive or destructive feedback on your performance. There's no symmetry to the conversation. Does that feel safe to you? I don't think that's safe.

"We have a set of core values that are important to us, and they're mostly around candor—really to generate speed, action orientation, and a sense of security. We've got twenty-one thousand people, so we have a lot of people who are managing others. What are the traits we want in leaders? We want managers to display confidence and optimism, and to give constructive feedback, never destructive. And managers need to seek constructive feedback themselves.

"The biggest organizational challenge I've seen in small, medium, and large companies is this issue of defensiveness. I'm mowing your grass, and maybe I'm making you defensive through my line of inquiry, or because what I'm doing overlaps with what you do. It creates defensiveness in the system, and it's a natural reptilian kind of response. That defensiveness is what over-amps corporate cultures. So you try to get defensiveness out of the system so that people are focused on achieving, learning, and bonding.

"And that doesn't mean that we have to go out to dinner, or go bowling. The point is 'Can I really take an interest in you, and you in me,' because it's meant to drive out the defensiveness that's part of so many conversations. People want to achieve. People want to learn. Generally, people are driven to do pretty constructive things. People really want to bond, but there can be so many defense mechanisms in a corporate environment. So I try to break that down, so that they feel they're in a safe environment. It's really hard on organizations when people in power throw their weight around. It creates an unsafe environment for collaboration. It breeds defensiveness."

Even symbolic gestures can help set the tone that employees' opinions matter at least as much as those of the bosses.

At the consulting firm QuestBack, cofounder Ivar Kroghrud, who was the firm's CEO for thirteen years before becoming its lead strategist, takes an unconventional approach with his company's organizational chart.

"I draw our organizational map upside down, because it's not the leader and manager who do the work," he said. "The manager is there to give direction and make it possible for the others to do the job. And that's clearly illustrated if you actually turn it on its head. Imagine showing a frontline employee the chart and saying, 'Let's find your box somewhere very, very far down here.' Just the psychology of that is very depressing because those are the people who deal with customers in a lot of cases. And that's a very important role."

Arkadi Kuhlmann, the former CEO of the online bank ING Direct until it was acquired by Capital One, took the unusual step of asking his employees to vote on whether he should continue as their CEO.

"All my colleagues think I'm nuts, and the board thinks I'm nuts. But I don't want to serve here unless I've got the commitment of people genuinely wanting me to serve," he said. "The vote is anonymous, of course. I'm not asking for a popularity contest. Part of it is, Do you have faith in the mission? Do you have faith in the company? Do you have faith in me? Now, the shareholders are okay with me, the board's okay with me, the regulators are okay with me, and my customers seem to like me. But what about the associates? It's a big question.

"The difficulty is getting people to interpret why I do this vote. I want people to get two things from this. One is that I don't take the job for granted. And, number two, that I'm willing to be accountable to them, not because I work for them in a broader sense, but I've got to walk the talk, right? So if I keep walking around saying all the time that our associates are so important, then why don't they have a say in terms of whether or not I'm leading?"

Banishing Yelling

Why do some people scream and yell at work? Perhaps they're under tremendous pressure—Must. Deliver. Results. Now.—and so they lash out in frustration at their underlings to create a sense of urgency or to send the message "Never do that again."

Bob Brennan, formerly of Iron Mountain, offered an insightful analysis of the kind of people who turned out to be yellers at the office—until he moved them out.

"Invariably, it's about hierarchy. They're most interested in being the boss," he said. "They're not trying to lead a team to find the right answer, the better answer, and to be open and vulnerable. So there's an alpha dimension to their personality that comes with the package. But if they don't have this ability to subordinate themselves to what we are trying to achieve, and what they need their area to achieve, that's where it breaks down for me. So when I fail with a hire, that's the dimension I fail with. Sometimes people are very good at carrying themselves in a way that you wouldn't think that, but when they are put in a position of distress, it reverts more to 'No, you will do as I say.' But it's really only under distress or opportunism. It's not in the normal course of events that those behaviors exhibit themselves.

"I'm looking for leaders who can step back and help guide those who are in the competitive moment. A lot of times there's a breakdown between the intent somebody has and the impact they have. Their intent is to really help you succeed. Their impact is to make you very defensive. They don't understand how they're coming across, which might be attacking and interrogating, and pushing as opposed to pulling. Some people might euphemistically refer to that as passion, and I say, 'Malarkey.' It's not passion. I don't believe that there's any room in business at all for yelling. And some people hide that behind this veil of passion. It's bad behavior, and has the wrong impact. You cannot lose it, ever."

That rule—you cannot lose it, ever—is ironclad at other companies, too. Robert Johnson of the RLJ Companies said he doesn't understand the purpose of yelling in business.

"I just don't want people to get angry. I just will not get angry, and I won't let people get angry and lash out at somebody personally," he said. "I just don't understand anger and conflict in a business. If you think about it, in a business you're working to make money for somebody—either yourself or your shareholders or somebody. How could you get angry about trying to figure out how to make money? If we're not angry, and we work together, we make more money. If we get angry and we have conflict, we make less money. So let's not get angry. Let's just work it out. It's not like if you get angry with somebody, all of a sudden the stock is going to jump twenty percent. Either you've got a business or a product that people are going to buy, or they're not going to buy.

"I start with the notion that if there is something that's broken, I can fix it. So there's no reason to get mad at somebody. I'm always starting with a certain amount of trust for anybody I'm dealing with. And by the way, even if you do get angry, it's not going to solve the problem. All it's going to do is reverberate around the office that so-and-so made a mistake and so-and-so is angry at them. Then a whole cloud of frustration and anger pervades the office. All of a sudden you get a breakdown in the culture of cooperation and collegiality, and the common mission goes out the window. And it'll take you a week or so to get everybody back together.

"Some people think I'm sort of not passionate or I'm kind of cold or disinterested because I don't rant and rave and everything else. I don't do that. I think it's a simple rule: More insecure, more anger. More secure, less anger. I think really great companies are populated by people who are confident, secure, and less fearful. Just think about companies that really stay at the top all the time. They don't have a lot of turnover. There's a lot of continuity because the environment is conducive to people wanting to be there, and they want to stay there."

The chef Mario Batali said that he has a no-yelling rule in his kitchens. His rationale is not unique to the restaurant business:

"One of the big rules for our kitchens is that if you're not close enough to be able to touch me, you can't talk to me. A lot of people will yell across the kitchen because it's just easier and faster. That doesn't work with us, so our kitchens are smaller, and you need to talk in a conversational tone. If you can't, you have to move toward me, because if you're yelling at me, then there can be problems understanding the nature of your message," he said.

"I worked with a lot of yellers over the years. My opinion is that yelling is the result of the dismay you feel when you realize you have not done your own job. Everyone in the restaurant business knows it's not going to be busy at five p.m. It's going to be really busy between seven thirty and nine thirty or ten o'clock, and then it's going to taper off a little bit. And it is as inevitable as Christmas. So it's the chef's job to prepare the staff for what will inevitably come. And it comes every night, so it's not like, 'Oh my God. What happened today?' The reason the chef yells is because the chef is expressing dissatisfaction with himself or herself for not having prepared you properly. And then, of course, the obvious scapegoat is the person who's the least prepared."

Dan Schneider, the CEO of SIB Development and Consulting, which helps companies find cost savings on their contracts, uses a surprising approach to handle big problems at his company. It involves ice cream.

"Years earlier, I used to scream and yell and throw things," he said. "I was hotheaded and it was stressful, and it was the same thing with my next company, because that's what worked before. I was your stereotypical lunatic CEO, and it wasn't like I was trying to be that way. That was natural.

"So when I started SIB, I didn't know if I was going to have that many employees or whatever. But I'd become more laid-back and relaxed over two and a half years of taking a sabbatical. I just handled things differently. I don't get upset. I don't really yell anymore because it accomplishes nothing. I couldn't say the one turning point that made me realize it. But there were moments. We had an employee who was working on a project for a Fortune 5 company, a major project. He'd been working on it for a month, compiling data, and he didn't back up his hard drive. So it's a month's worth of work, tons of data. We were able to get the hard drive recovered but it cost about fifteen hundred bucks. Everybody looked at me like they thought I was just going to fire him or kill him.

"And I don't know where I got the idea, but I just said: 'You've got to go buy everybody ice cream. We're having an ice cream party.' Because I realized that if I yell at everybody, they're just going to figure I'm a jerk. But if we're all sitting around eating ice cream, everybody knows why we're eating ice cream—it's because this guy screwed up. That will set in and they'll remember it and then just maybe they'll think, 'Oh, yeah. We had ice cream last week. Maybe we should back up our work.' Knock on wood, we haven't had to recover any more hard drives."

Playing to Strengths

Paul Maritz, the former CEO of the software firm VMware, said leaders should play to people's strengths, not their weaknesses. Taking a more positive approach in this way will help CEOs build up a store of goodwill with their employees that will come in handy when it's time to shift direction quickly.

"It's very hard to talk about these things without becoming trite or corny, but the best leaders are those who get the best out of other people," Maritz said. "I've learned that you only really get the best out of other people when you do things in a positive way. There are negative styles of leadership, where you do things by critiquing and criticizing and terrifying other people. But in the final analysis, it doesn't get the best out of people and it doesn't breed loyalty. Because no matter how much we think we've got things figured out, we haven't got things figured out. Inevitably, we're going to go down blind alleys. We're going to run into problems. We're going to make mistakes. And when that happens, you have to ask people to help you and to overlook the fact that you've messed something up.

"Great leaders, in my view, are those who have built up that reservoir of loyalty, so that when the time comes to say to folks, 'We have to change direction,' people are willing to make an extraordinary effort. If you're the kind of leader who cuts people down and humiliates them, you leave scars on people that can eventually come back to haunt you."

Because of some scars left by an early boss, Irwin D. Simon, the CEO of Hain Celestial Group, a natural and organic food and personal care products company, said he developed a leadership philosophy of building people up by, as he calls it, "spoon-feeding" them.

"I am somebody who has learned throughout my career about empowering people, about how I don't have to be in control," he said. "It's about not having an ego out there. We all have egos, but don't let ego get in the way. I don't have to show people I'm the boss or I'm the leader. Just by treating people right, I find that they want to be part of your team.

"There were a couple of important inflection points for me. Number one, I worked in some corporate environments that were very political. If you got behind the right people, you would do well. And if you didn't support certain people, you were off the team—your competency or loyalty didn't matter. You just didn't make it. So I'm a big believer in the idea that we're all on the team together, and you have to treat everybody equally. It's not the select few—'Here's the boys' club, here's the girls' club.' I'm a big believer in bringing everybody in.

"The other really big one in my career happened when I was working for a certain gentleman, and I used to ask a lot of questions. One day he pulled me aside and said, 'You're asking too many questions, and you are perceived as smarter than me, and I'm the chairman. And you shouldn't be perceived as smarter than me.' I love to be around smart, fun people. And if you're confident, let your people ask questions, and do things, and speak up.

"I'm consistent at work. It's not 'Monday I mean it, Tuesday I don't.' It's just been the way from the beginning here. It comes back to hiring good people who are smart. You're not going to embarrass them. The point is to make them feel good. There were some situations in my career where some nasty things were said to me. There's scar tissue from that stuff. I think it's important, with people who work with me, to spoon-feed them. Spoon-feeding them means playing to their strengths, to help them build confidence. It's not just my senior executives. It's everybody. You should know what people want because I know what I want. I know how I like to be treated. And you just take that and say, 'How do I treat people in the same way?'"

Respect is a powerful and necessary tool for a leader who hopes to get the best work out of people. Jeffrey Katzenberg, the CEO of DreamWorks Animation, put a particularly fine point on this idea.

"By definition if there's leadership, it means there are followers, and you're only as good as the followers," he observed. "I believe the quality of the followers is in direct correlation to the respect you hold them in. It's not how much they respect you that is most important. It's actually how much you respect them. It's everything."

5.

IT'S ABOUT THE TEAM

Check your ego and your title at the door.

—Enrique Salem,
former CEO of Symantec

In the previous chapter, several CEOs made a compelling case for the crucial role of respect in effective cultures. But respect is just part of the equation; performance and accountability also matter. For any company to operate at a high level, people have to play their positions.

Companies are like teams, and for teams to succeed—especially as the speed of business picks up—all the players have to deliver, so that their colleagues can count on them. Are you dependable? Do you recognize your role as part of the whole? Can people count on you to carry the ball, not drop it, and even anticipate plays so that you're in the right position when it matters? Do you take the initiative to go after the ball, and not simply wait for it to come to you?

Call this characteristic trustworthiness. Call it dependability. What it means is that you recognize your role on the team and play your position. When everybody does that, the team can focus energy on executing the strategy, rather than being distracted by worries over whether the teammates are going to do what they're supposed to do. When companies can find the right balance—treating each other with respect, while also setting clear expectations that everyone must play his or her role—the group becomes greater than the sum of its parts.

It's a lesson that Peter Löscher, the CEO of Siemens, learned as the captain of his volleyball team in high school and in college.

"At the end of the day, it's about fostering the best performance from the people on the team," he said. "It's less a question of how you train and your physical conditioning. The difference between a good team and a great team is usually mind-set. When you watch great games in sports, you see there's a moment, all of a sudden, when the team clicks. It's something that's always caught my attention—why and how that happens with teams.

"When you're in business, I think the underlying principle is trust. How do you establish within a team a blind trust so that each person plays for the other? Business is about lining up a leadership team or a group of people and you rally them behind a cause or a certain direction. But the underlying strength is the trust within the team, so that you actually are no longer just playing individually at your best, but you're also trying to understand what you can do to make the team better."

Do What You Say

If you think back on many of the colleagues you've worked with over the years, you can probably divide them into two broad categories. You could rely on some people to do what they said they were going to do—they always answered e-mails, they followed up, and nothing fell through the cracks. Now think about everyone else, and about all the moments when you said to yourself, "What ever happened with that? I need to ask them about that." Moments like that, multiplied across an entire organization, add up to a lot of wasted energy and lost momentum. To counteract this, many CEOs establish a simple rule for their employees: they have to do what they say they are going to do.

In an earlier chapter, Steve Stoute of Translation LLC and Carol's Daughter explained that "doing what you say" is one of the core values of his organizations. Here he explains its importance:

"I'm very up-front, I expect the best, and I hold people accountable for everything that comes out of their mouths. Don't say you're going to do something and not do it, because in a company of this size, everybody is directly responsible for the person next to them.

"It's like one of those moments where everybody's holding hands. So if somebody doesn't do something, it's felt throughout the organization. The organization's not big enough to withstand those kinds of errors. At big companies, that happens all the time, and it can take years before it starts to affect the bottom line. Small organizations have the benefit of being nimble, but the threat is that when one person catches a cold, everybody catches a cold."

Shellye Archambeau of MetricStream uses a simple metaphor— "Who's got the ball?"—to signal that she expects people to deliver on their responsibilities:

"When you're in sports, and the ball is thrown to you, then you've got the ball, and you're now in control of what happens next. Which means you own it. So it's important to know who's got the ball. If you're in a meeting and you've had a great conversation and then everybody leaves, who has the ball? It becomes a very visible concept for making sure that there's actually ownership to make sure things get done. And it's one thing if you always catch the ball if people toss it to you. It's another thing if you are proactively going after that ball. As leaders, you've got to make sure that you're actually going after that ball."

Tim Bucher, the former CEO of TastingRoom.com, a wine site, makes the case that trust is essential for speed in a startup. But his insights apply just as well to large organizations.

"I learned that one of the most important leadership tools— again, for start-up life—is an expression that I tell my team every time we have an all-hands meeting: 'Trust until trust is broken.' Because when you trust until trust is broken, you can move really fast," he said. "I will tell a new executive on the team when I hire them: 'From day one, you have my complete trust. I trust you to run your part of the business the way you deem appropriate. And I'm going to support you in that, and you're going to make decisions and I'm going to back you. I'm going to do that until trust is broken, and hopefully trust is never broken.'

"You can move really fast as an organization with that philosophy, especially if you can get that philosophy throughout the whole organization. When you instill that from day one in a company, it's pretty powerful, because then everyone's got everyone else's back."

Trust means a firm belief in another person's reliability, truth, ability, or strength. But reliability is key. If people do what they say they're going to do in an organization, then that builds trust, and trust builds teamwork and speed.

Enforcing the "No Jerks" Rule

Enrique Salem, the former CEO of Symantec, said football taught him an important lesson about playing his position on a team.

"I was captain of the varsity football team my senior year of high school," he recalled. "We called the plays the coach would signal to us from the sideline. I used to be very much a student of the game. I would watch the game films myself and get ideas of what we should do, what we should think about.

"One time the coach called a defensive play and I changed it, and after having some success with that, I said, 'Oh, this isn't so hard.' But then another player runs on the field and replaces me, and I run to the bench and the coach says, 'When you want to call what I'm calling, you can go back in the game.' So I sat on the bench for a play or two and then went over and said: 'Okay, Coach. I got it. I'm sorry.' And he put me back in the game. I really learned this notion that whoever's making the calls, you've got to listen to that person.

"He pulled me aside after the game and we talked about it, and he said: 'I know you love the game. I know you study the game. But you've got to realize that when I make calls, I'm setting something up. I'm looking at something that's happening, and you can't be out there second-guessing me on this.' You run into situations where there's a bigger picture sometimes that an individual who's working on a project may not be able to see, and can't understand all the implications of any decision you make.

"I now have this presentation that I give to our advanced leadership class, and the title is 'Lessons I've Learned Along the Way.' The very first chart says, 'Check your ego and your title at the door.' I learned that very early on. One of the things that my first manager said to me was 'Look, a lot of times you don't lead by your position. You lead by how you influence other people's thinking.' And so I absolutely believe that if it's about you, you're not going to do a great job. It can't be about your success. It has to be about what you are trying to accomplish. So that's number one."

That is why so many CEOs have adopted some variation of the "no jerks" rule for hiring. They need people who are team players, first and foremost. Michael Lebowitz, the CEO of the digital marketing agency Big Spaceship, said he had learned the hard way to establish a culture that doesn't tolerate "rock stars."

"Probably the biggest lesson I learned as we started to grow was—and this is a more sanitized version of the expression we use—'Don't hire jerks, no matter how talented,'" he said. "I became very attuned to this early on, when we were still a small start-up, and you're doing everything you can to maintain a positive framework. So I'm looking for people I like, because I've seen how, no matter how talented they are, the negative is always going to pull down any positive. The second- or third- or fourth-best candidate who isn't a jerk is going to ultimately provide way more value. Because we learned that early on, we've always guarded against that sort of rock star culture.

"People like that can say all the right things in interviews, and then they come in and really make people's lives miserable. You spend at least a third of your life at your job. You should have a place you're happy to go to every day. And if you're not making good on that in even the smallest way, it becomes sort of pernicious. It can amplify itself very quickly.

"I remember a guy, he really was an incredibly talented designer, one of the best I've ever seen, but he was just surly. No matter how good you are, design is always an exercise in balancing what you think is best with someone else's needs, even arbitrary things. He couldn't roll with that stuff. He had conviction born of great talent, but he was in the wrong business to have that kind of attitude. He was mostly battling with me, but I think it kind of gave permission for that attitude—almost invited it—for the other designers. They felt resentful that I was paying that much attention to that person rather than just sort of saying, 'What are you doing?,' which I should have done. I was treating him like a rock star, fundamentally. And I've done that a number of times since and each time I realize it and I have to put a stop to it because that won't play in the kind of environment that I want to create. And so you can't hire the rock star. It really is damaging."

Chauncey C. Mayfield, the CEO of MayfieldGentry Realty

Advisors, said that some early experiences affected his decision to create a culture where the work itself was more important than individual egos.

"I once worked in a company that, walking in the door, it seemed like everybody was given a big club to see whose knees we could knock out from under them first," he said. "I was pretty good at it, but I said I never want to work in this kind of environment again. It had to be fun, and I didn't want it to be simply a pressure cooker. So if you can't have fun here, then why are we spending all this time here? And what I mean by 'fun' is there are a couple of very basic principles that we operate with. One is that, when you come to work, and you have an assignment that you absolutely cannot complete, you're not penalized for saying, 'I'm not sure if I can do this.' We encourage you to do that. You're penalized if you don't tell us you can't do it. We'll find someone to work with you to do that.

"There are some things I do extremely well and some things I am not good at doing. I think what happens in some companies is that people are put in positions where they can't complete the assignments. But they're concerned about what is going to be said on their performance review and how it's going to affect their bonus. So all of a sudden they sit there and they sort of dabble and worry over it. And they dabble over it and they eventually turn in something and it leads to the same result, which is a bad performance review and they probably won't get moved up as quickly. It shouldn't be an environment where you come and worry all the time. That's not what I want for us.

"The second thing is that it's got to be a team environment. We, first of all, pick the best team leader. And oftentimes, the best team leader is not the most senior person. The objective is to win. So you can have a person who's at a pretty junior level leading the team with senior people on it. Now, what that requires you to do is to check your ego, because your middle manager is now giving out assignments. And for the partner, your assignment may not be the assignment you like, but you know what? You're on his team or her team. The point is, we're all smart, that's a given. But are we smart enough to take our egos and park them? So when I'm hiring, I'm trying to figure out 'Are you able to play that third position on the team?' Because we need you to play third position. And if you are, then it's probably a pretty good fit for us."

Niraj Shah, the CEO of Wayfair.com, which sells home furnishings, said he learned a memorable lesson from a mentor about putting the team first, even when people might disagree with a decision about strategy.

"Somebody I worked for about twelve years ago had a big influence on me. His general management sense and skills were quite amazing. In any given situation, it was always pretty clear to him what to do. He had a real good sense of how you manage people and what makes sense and what doesn't make sense and how you build a team," he said.

"There was one situation I remember in particular. Another executive in the company had a plan for something, and I didn't think the plan made a lot of sense. I was quite frustrated that we were pursuing that plan because I considered it a waste, and I think my manager agreed with me on that particular situation. I was going on about how this doesn't make sense, and that we were going to waste a bunch of money and not accomplish our goal. And he shut the door and said, 'You need to support him and support the plan.' And I said, 'Well, why would you do that?'

"And he said that there is a time you debate, but once a decision is made, you need to support the person. Because the day you have somebody who works for you and you're not supporting them should be the day you fire them, because they are never going to be successful. He said you need to act as a team, and there will be times you disagree and there will be time for healthy debate. But if you make a decision, then you have to all pull in the same direction."

John Donovan, the chief technology officer at AT&T, said his approach to work changed when he learned to stop worrying about whether he was getting credit for successes.

"I developed team skills because I started to engage in deliberate deflection of credit in an environment where it was all about credits," he recalled. "What I started realizing is that people appreciated when you played for the result, and not for your role on the team. So I learned there that giving credit away, deflecting credit, was an effective thing to do. I think I learned a lot of subtleties about teams and how you assemble teams.

"If you figure there's a karma pool out there floating around for credits, you have to stop playing for credits. I remember the day I realized that, and I probably never again needed to involve score-keeping in anything that I did."

The Team at the Top

If the leaders of companies and their executive teams expect all employees to work together effectively and play their positions, they have to be role models themselves. Many CEOs said they spent considerable time deepening a sense of trust and clarity about how they will work together as a team.

Laura Ching, a cofounder and the chief merchandising officer of TinyPrints.com, an online card and stationery company, recalled having those detailed discussions with her two partners as they were starting the company.

"The backdrop was that we all were about four years out of business school or working at big companies, and we kind of got that entrepreneurial itch," she said. "We are very competitive, driven people and wanted to try to build something great, but we were also motivated about creating a culture, because a lot of things were moving too slowly in the jobs we had. And we didn't have the idea at first for the kind of business we would start. We just knew that we all had similar values. And so we spent about six months—a bunch of us in a tiny apartment, every Wednesday night, over burritos—brainstorming all these crazy ideas.

"We didn't know where it was going to go. But we spent a lot of time just forming our partnership agreement. What were our expectations of each other? How long did we expect each other to be committed? What happens if someone's not doing a job right? And I think that honesty and candor really were a foundation for some of the values that we wanted to set. I have to say that in the beginning, it was like, 'Come on, let's just get started.' We don't even know if this is going to mean anything. We didn't have a business plan. We spent more time saying, 'Okay, what's the ownership going to look like?' But I am so glad we did it because there have been situations that have come up between us where I'm glad we already thought about them early on. It's easier to agree on something when everyone's being objective."

Deborah Farrington, a general partner at the venture capital firm StarVest Partners, went through a similar exercise with her cofounders.

"There were four of us who started the business together— three women and one man—and we had each worked at firms that we felt had been too aggressive, too cutthroat," she said. "We came from investment banks, from private equity, from other venture firms, where people were always vying to say, 'I did this, and therefore I want the biggest reward.' We wanted to start a firm that was based on respect for the individual, with a moral culture. So we started out with a rule: 'No jerks allowed, ever.' The language was a little more blunt, but you get the idea. We'd all worked with too many people like that.

"And we wanted a culture where accomplishments and results were valued above politics. Each person would have time to be heard. And each person would be required to be heard. You had to have an opinion. You couldn't just say, 'Well, I'm going to pass on that.' Another thing we did was to bring in a leadership consultant not long after we started the business in 1998 to run off-site meetings for the four founders once or twice a year. He helped us understand our individual styles and how we liked to make decisions and interact. One person went A, B, C, D—very linear. Another person quickly processed things, immediately went from A to Z, and said, 'This is it.' And then other people had something in between. So it really helped us understand the decision-making style of the partners, and that helped us develop some of the ground rules for how we make decisions. So we really focused on communication and clarity. One of our rules is that everybody gets a chance to speak, and they have three minutes each. Then we would discuss, and then we would vote. We really focused on the process."

The CEOs of start-ups weren't the only ones who shared such stories. Caryl M. Stern described how she developed a culture of teamwork when she took over as CEO of the U.S. Fund for UNICEF.

"It was a very interesting time in my life," she recalled. "I had taught leadership development at Manhattanville College. So it was a chance to take all of this textbook learning and actually apply it. Can you do it? Can you really have a work team? And I've never worked with a better team than I'm working with right now, and I've never worked in an environment as energized as the one I'm working in right now.

"And that didn't happen by chance. We hired coaches to help make that happen. We wrote values to help make that happen. We decided we wanted to be the nonprofit you'd want to work for.

We had a staff retreat and we did a blowup of a magazine cover with the senior management team on it that said, 'U.S. Fund for UNICEF Named Charity of the Year Five Years Out.' And we spent a weekend holed up in a hotel, and we wrote the article. If we were going to be named the charity of the year five years from now, what would we have done? What would we have accomplished? And we spent a lot of time on that, but also, What would we be internally? What would it feel like? What would you as an employee expect? What would I, as a boss, want from you? What's the environment? So it wasn't only about what will we achieve, but how are we going to get there?

"We hired a coach who worked with us collectively but also coached us individually about process—not skills, process. He actually took us through the process of learning how to work together, and it was the most phenomenal thing I've ever been a part of. We have an entire team really working together now, so the sum is so much greater than its parts. I've never sat at a management table before where people will say, 'Well, my division really needs this, but I think your division's needs are bigger, so I'm going to put mine off to the side.' I'm not describing utopia. It's a competitive environment, but there is a sense of trust and a bond within this group.

"You have to take time to get to know each other—not to be friends necessarily. If you end up as friends, that's great. But I need to understand what you need to work, and you need to understand what I need to work. If I understand that time has no meaning for you and time has a lot of meaning for me, then we'd better negotiate how we're going to handle that. And oftentimes in management experiences, we don't have those discussions. So you come late to meetings. I show up on time. The meeting now starts fifteen minutes late. I'm angry; you're angry. And this is how we're going to do our best work? No, this doesn't really work. So I need to sit down and negotiate with you. But we are going to honor that the six people on our team are the most important six people to each other. So that means if you expect something from me by five o'clock, you're going to get it. You have to be able to do your work all day and not worry about whether that's going to be in your inbox by five. And the moment you have to worry about it, the team doesn't work."

Bill Flemming, the president of Skanska USA Building, also hired a facilitator to help improve teamwork among the top executives.

"Teamwork is key in this business," he said. "This is not an individual sport. I see many leaders who are somewhat egotistical. To me, it's more about the team. And in my early years in this job, I focused on organizing the senior leadership so that they're moving more in the same direction. The first step was acknowledging that I wasn't going to do it. I'm part of the team, not the guy who's going to lead everybody in how to change it. I realized we needed a facilitator to do that. So I brought somebody in just to teach us better interaction.

"The questions he asked everybody in the group included 'Why do you want to be part of this team?' 'Why do you want to be in this company?' 'Why do you show up at work?' 'What's in your box?' And when I say, 'What's in your box?' it means 'What drives you?' and I don't want to hear about the party line that you're doing this for the company. And 'What commitment do you have to your partners?' We each had to come up with a personal commitment to our team and talk about our responsibility to others on the team. That drove people together fairly quickly. The interaction in the group has been different."

Paul Maritz of VMware offered an intriguing analysis of the essential components of any executive team. For the team to work together effectively, he said, people need a collective self-awareness of the group.

"One of the things I've learned over the years is that there is no such thing as a perfect leader," he said. "If you look at successful groups, inevitably there's an amalgam of personalities that really enable the group to function at a high level. At the risk of oversimplifying, I think that in any great leadership team, you find at least four personalities, and you never find all four of those personalities in a single person.

"You need to have somebody who is a strategist or visionary, who sets the goals for where the organization needs to go. You need to have somebody who is the classic manager—somebody who takes care of the organization, in terms of making sure that everybody knows what they need to do and making sure that tasks are broken up into manageable actions and how they're going to be measured. You need a champion for the customer, because you are trying to translate your product into something that customers are going to pay for. So it's important to have somebody who understands how customers will see it. I've seen many endeavors fail because people weren't able to connect the strategy to the way the customers would see the issue.

"Then, lastly, you need the enforcer. You need somebody who says: 'We've stared at this issue long enough. We're not going to stare at it anymore. We're going to do something about it. We're going to make a decision. We're going to deal with whatever conflict we have.' You very rarely find more than two of those personalities in one person. I've never seen it. And really great teams are where you have a group of people who provide those functions and who respect each other and, equally important, both know who they are and who they are not. Often, I've seen people get into trouble when they think they're the strategist and they're not, or they think they're the decision maker and they're not. You need a degree of humility and self-awareness. Really great teams have team members who know who they are and who they're not, and they know when to get out of the way and let the other team members make their contribution."

Team, Not Family

You hear this from a lot of organizations: "We're a family." It's a nice notion that suggests how close people are—how they look out for each other and care about each other and their families. It's not just about the work.

But Linda Lausell Bryant, the executive director of Inwood House, a nonprofit that focuses on teenagers' health issues, has a strong opinion about this approach.

"Recently, I've really shifted my thinking," she said. "Our culture reflected our work, which is to create a sense of family for our teens. So our staff would say: 'We're a family. We're a family.' And I've actually said directly to everyone in all-staff meetings: 'We're not a family, because in a family you never can fire somebody like your uncle Joe. You just can't. You have to put up with him because he's family. In an organization, if someone is taking the organization down, we can't accept that because the organization is bigger than any one of us.'

"So I've said to them that the analogy that best suits us is 'We're a team,' and in a team, everybody's got a role to play. And the team wins when everybody plays their roles to their best ability. The other thing that's different in a team is that people understand the concept of roles. So if you're the manager, you have a job to do as a manager. No one, generally speaking, resents the fact that you have authority, because they understand that it comes with the role of a manager and that teams need managers. They don't manage themselves.

"But in a family, it is about power. You know, Mom or Dad has the power, and I think the dynamic that often plays out in a workplace is that people project all of their parental stuff. And I remember a job where I actually had to say to my team: 'I am not your mother. I'm the division director here. I have a job to do. You have a job to do.'"

6.

ADULT CONVERSATIONS

> *If I really care about someone, then it's better to be honest with them, to want them to succeed and to say the right things that will push them to be better.*
>
> —LAURA CHING, cofounder and
> chief merchandising officer,
> TinyPrints.com

Seth Besmertnik, the cofounder and CEO of Conductor, a search engine technology company, remembers vividly the first time he managed an employee. It did not go well.

"I started Conductor when I was twenty-three, and my first official management experience was with the first salesperson we hired. He was very difficult to manage, and I did everything wrong as an early manager. My second week working with this guy, he got a call from somebody who would be a great new customer for us. But the customer was in California and said he could only talk at six p.m. My salesperson came to me and said he couldn't do the call then because he had to go to the gym. My blood was boiling. But I was intimidated by this guy, and I didn't have the courage to actually say, 'What are you doing? This is terrible.'

"We probably should have fired the guy on the spot. That would be my inclination today. But eighteen months later, I had never once given this guy any kind of critical feedback. And I had my employees from the organization telling me that this guy's a problem. I ended up letting the guy go, and he was furious. He had every right to be furious, because I had never given him any critical feedback over eighteen months. I'd say he was my first real management experience, and I got an F."

But how many managers do much better?

In many companies, managers are afraid to offer frank feedback. As a result, problems are swept under the rug, tensions simmer, and talks that should have happened in the moment are delayed for months, until a performance review. At that point, they take on outsized importance because the employee, hearing about a problem for the first time, feels blindsided and resentful, gets his back up, and maybe even lashes out. The bad experience makes the manager even more reluctant to address a problem in the future.

"I call them adult conversations," said Geoffrey Canada, the CEO of the Harlem Children's Zone, a nonprofit. "People don't want to have these conversations. They will avoid them. I am often asking people: 'Do you need help with having these conversations? So you're going to talk to that person. What are you going to say? How are you going to say it?' It's something that most of us aren't trained to do. I used to think it was particularly true in a not-for-profit business, because people who come to work at a not-for-profit want to help people, so it's harder for them to make these tough calls. But the more I have seen people in for-profits, I think they're even worse at it.

"I'm just stunned sometimes with how unwilling people are to bring somebody in the office and just say to them: 'Look, you're a good person. You know I like you. I like your family. This job is not really working out, and I'm going to have to let you go.' And so they will put people in another position, where they don't really add to the bottom line, so they don't have to deal with firing them. We don't have the luxury of being able to do that. I mean, we don't have positions that we can just sort of stuff you in."

And when grudges build up, sometimes good people leave, said Conductor's Seth Besmertnik.

"I think employees often resign from companies because they had a problem with something, and all these little things fester. And they never once share them with anybody. Then they come in and say, 'I'm out of here. Here's my two weeks' notice.' Their manager will say, 'Why are you leaving?' And they'll answer, 'I'm upset about this, this, and this.' The company might be able to fix all those things, but it's too many conversations to unravel, and it's too late."

The simple approach described in the previous two chapters—that employees should treat one another with respect while also playing their own positions on the team and expecting others to do the same—can succeed only if staff members are willing to have frank discussions to work through the inevitable disagreements and misunderstandings.

To be sure, it's not easy. Many CEOs say they had to learn how to have adult conversations. But they also recognize that these conversations can uncork energy that is otherwise bottled up because people are reluctant to say what's really on their minds.

Starting at the Top

If leaders want their managers and employees to be frank with one another, they have to set an example themselves. Many CEOs say it is a simple test for effective leaders: Can you look someone in the eye and give them tough feedback?

"A lot of my growth as a manager has been around conquering

my own insecurity and gaining confidence," said Seth Besmertnik, of Conductor. "When you're confident, you can give people feedback. You can be candid. You feel secure enough to say what's really on your mind, to bring someone in the room and say, 'You did this. It really made me feel XYZ.' Having good conversations is really eighty percent of being an effective manager."

Laura Ching of TinyPrints.com described how she learned the importance of giving clear feedback to avoid misunderstandings.

"I worked at Walmart.com after business school," she recalled. "It was really challenging. I just remember thinking that it will be so easy to be a good boss because I had been frustrated with bosses before and wondered, 'Why don't they get it?' But once you start doing it, you learn so much.

"One of the things I learned was that it was really hard for me to give negative feedback. I'm kind of a people-pleaser. I like it when people are happy, at work and in their home life. And it took me a while to figure out that it's better to give that negative feedback right away. I remember one person I managed who was doing an okay job, and I gave her constructive criticism, but I balanced that with positive reinforcement, too. I think she only heard the positive, and the negative didn't resonate.

"By the time we got to the review period and she didn't get a great review, we were just so disconnected. And that was a horrible feeling for me. I wondered, What did I do wrong, and where did it all fall apart? It taught me that I needed to be more comfortable with giving that feedback. If I really care about someone, then it's better to be honest with them, to want them to succeed, and to say the right things that will push them to be better."

As difficult as adult conversations can be—especially when they include critical feedback—the impact can often be powerful if people learn something new about themselves in a constructive way.

Karen May, the vice president for people development at

Google, has coached hundreds of executives during her career, and she has often seen the benefits of such feedback.

"As a coach, I was often in the position of giving people feedback that they hadn't heard before, after interviewing a bunch of people they work with," she said. "It was always difficult for me, too. Just at a human level, it's difficult to tell somebody something that isn't working about them. But I came to find that people are incredibly grateful. If I'm not doing well and I don't know it or I don't know why or I can't put my finger on what's not working and no one will tell me, then I don't know how to fix it. And if you give me the information, the moment that the information is being transferred is painful, but then I have the opportunity to change it. I've come to realize one of the most valuable things I could do for somebody is tell them exactly what nobody else had told them before."

For Brent Saunders, the former CEO of Bausch & Lomb, being direct was a skill he had to learn quickly early in his career, when he worked in the high-stakes world of consulting.

"When I first started managing people, you wanted to always just pat them on the back and say, 'Great job,' and when they did something wrong, you wanted to hide under the desk and pretend it never happened," he said. "But one of the things I learned was that, particularly when you're selling human capital and brain power, if you didn't deal with those things right away and directly, they could turn into a bigger problem because the work could be bad, and people are paying by the hour, and so a redo would be horrible for a client in those situations. So I figured out very quickly by some trial and error that you had to deal with those things right up front. That's something that stayed with me throughout my career."

Linda Lausell Bryant of Inwood House said her background in conflict resolution training not only made her comfortable with tough conversations but also made her appreciate their importance.

"There are understandable tendencies among people to say, 'Let's avoid conflict,'" she said. "I was trained to really go for it, and find out what some disagreement is about. What's really underlying it? What are the underlying needs and issues here? So two people will present the conflict as 'I wanted the red one; she wanted the blue.' Or whatever it is. But is it really about the red or the blue, or what's it really about? I've always felt particularly adept at finding out the underlying psychodynamic issues. The training to not avoid the conflict—to kind of go for it and learn to get comfortable with it—was something that shaped me very much.

"You have to respect not only people's needs, but also their pain, their vulnerability. A lot of battles are about very personal things. I'm very attuned to the unspoken needs that people play out in the workplace. People are people in whatever setting—they bring their luggage of stuff; we all do—and the dynamics in the workplace are a function of the interaction of what we all have in our suitcases. You can't change that. You can acknowledge it. You can give it space. You can give it air and light. In the end, it can't rule the day, either, because in the workplace there are higher things and rules that are going to guide what we need to do here. It's helpful to know that, and be aware of it as a boss, and it's even better if employees are aware of it and that they feel that you're not trying to change who they are."

Giving Feedback

At Proteus Digital Health, Andrew Thompson has established groups to work on different aspects of the company's culture. One group focuses strictly on feedback, teaching people how to talk to one another about their performance in a respectful way. The group uses the memorable image of a tennis player going "over the net" as a reminder about the rules for giving feedback.

"One challenge we wanted to address," Thompson said, "is that we wanted to be a company where people could talk to each other honestly and give each other feedback directly rather than letting it fester or going to their boss and saying 'I can't work with Fred because he's whatever.' It's much better to go to Fred and say, 'Hey, look, when you do X it makes me feel Y.' So everyone in the company now has had feedback training.

"As the CEO, that meant that one of the first things that had to happen was that I participated with the management group, which is a fairly large group of about thirty or forty people. You sit in a big circle, and they can all give me feedback. Actually, it's a very positive experience because people have been trained so they know how to give it in a way that's appropriate and balanced so they can get their message across. So now everyone does that. We're still not that good at it, but we're learning and we're trying.

"You've got to have people understand how to talk to each other, and it's got to be direct. It's got to be in the moment, and it can't be 'over the net.' When you give somebody feedback, it can't be to say, 'You're doing this because you don't like me,' or whatever. It's got to be a very straightforward thing where you say, 'When you yell at me, it makes me feel like I'm not valued.'

"If you're over the net, that means that rather than describing the behavior and how it makes you feel, you start explaining to the other person what their motivations are for their own behavior. That's where you get so many problems, because people see the behavior and, rather than giving feedback, they sit there and stew and concoct all the reasons why it's happening. People concoct all this imaginary garbage about why the person is doing this to them when in fact the person may not even realize that he is doing anything. It's like in tennis or volleyball, and you have to stay on your side of the net. And it's very simple, right?

"What I find interesting in these environments is understanding fairly specifically how what you may do or how you interact affects different people very differently. In one example, there were folks who felt that my style was just very intimidating and very demanding, and other people who found it very productive and very rewarding. And some of it is about understanding and explaining what I'm doing and why I'm doing it. Human systems are very complicated, particularly as you gain any type of scale in an organization. But it's important, because the culture that you build and the ability to be adaptive will define the quality of the outcome."

At Conductor, Seth Besmertnik adopted another strategy of giving constant feedback to employees:

"Every employee who joins the company gets a book called *Fierce Conversations* [by Susan Scott] and a letter I've written that basically says, 'Life moves forward one conversation at a time. If you can have effective communication here at the company and if you can learn how to have hard conversations with people, then that's going to solve most of the problems that come from work experience.'

"One approach I use is that when I first start working with someone, I say, 'Hey, I really like to give you feedback. It's my job to give you feedback.' Set their expectation that you're going to be giving them a lot of feedback. When they do something great, send them a little e-mail. When they do something you think they can do better, pull them aside, and get into that rhythm. A lot of bad patterns happen when you go for really long periods without giving people feedback, and it just bottles up. They're so used to not getting any feedback, that when they get it, it's this huge deal.

If you get into a rhythm of giving feedback, they get kind of used to it and desensitized to it. So I'm always reinforcing these cultural points. If you've got something on your mind, let it out. Don't let it drift."

David Rock of the NeuroLeadership Institute says that because people naturally feel threatened by critical feedback, another strategy that can work in many situations is to have people provide their own feedback. That makes them feel more in control, and elevates their status in their own mind, rather than feeling their status has been diminished by their supervisor.

"Usually we can predict that the person who's receiving the feedback is going to argue," he said. "And once you've tried it a few times as a manager to give feedback, you see people really arguing and pushing back, and what we think will be a five-minute conversation ends up being an hour and a half, and you find yourself going around in circles. That's exhausting, and that's a threat response. People will defend themselves. You can see it in their body language.

"Our research shows that about seventy-five percent of the time you can get people to give themselves feedback, and we call it self-directed 'feed forward' rather than feedback. So you can get people to give themselves feedback, which is actually a status reward for them, rather than a status threat.

"Let's say you've just blown a client meeting and I'm your boss and I know the meeting's gone badly. If I say to you, 'That meeting went badly. What went wrong?' then you're going to defend yourself. You're going to feel the status attack, and all your cognitive resources will go toward defending yourself. But what if I say to you, 'You're a smart person. I bet you've been thinking about that meeting. What are your thoughts on what you'll do next time?' Then I'm giving you a chance to look good, and you'll now reflect and think deeply about what you might do next time. If there's not a strong threat, and you're not fighting against something, it does turn out to be intrinsically rewarding for people to talk about how they might do something better next time. With this self-directed approach, most of the time you actually get to good insights and useful ways forward."

Dominic Orr of Aruba Networks said he uses a simple strategy—acknowledging that everyone is entitled to be "momentarily stupid"—to help frame conversations and disarm people when he's giving them feedback.

"You try to be intellectually honest with yourself, meaning that you have to forget about all the face-saving issues and so on," he said. "I tell people that if you work for me, you have to have a thick skin because there's no time to posture. I also tell people that everybody can be and will be momentarily stupid. I think that in many large companies, a lot of politics arise because somebody makes a statement in a meeting, and then it's weeks of wasted time and effort because they have to dig in to defend that position, and then politics come into play because they now want to lobby for their position.

"So when I interview key executives of my staff, I tell them that they need to accept that they can be, and will be, momentarily stupid. If they can accept that and be able to say, 'Oh, I was momentarily stupid; let's move on,' then you don't waste time dealing with that. I try to set an example and to be very thick-skinned. I have a very open door. I encourage a lot of feedback so that my staff has no inhibitions to just tell me that I was momentarily stupid or I actually was wrong in some way. Sometimes I argue. Sometimes in the end I'll say, 'I still want to go with my hunch.' I think I would fail in this whole management philosophy if my staff couldn't be intellectually honest with me. That's one principle I try very, very hard to set by example."

Orr elaborated on how he uses this approach to give people feedback:

"I sometimes write an e-mail to someone, and will add a section that begins, 'Start of intellectual honesty moment,' and then I will be just doubly hard on them, and then I will write, 'End moment,' and then I continue the e-mail. So you create, really, a little space for people. The point is to be very honest, and I try to do it one-on-one so they save face. The major thing you want to accomplish is to not make it personal. Then people will feel that you're not attacking them. You're just attacking the issue, the fact that he was behaving a certain way and you make it very kind of private."

A little humor can go a long way in disarming people, too. Tim Bucher of TastingRoom.com will sometimes invoke his late grandmother to deliver feedback.

"If you worked for me, I would never tell you that you did a terrible job," he said. "I think being direct is important, but you can capture it in a different way, and here's how I do it: I'll say, 'I think my grandmother could have done a better job on that.' It's a really simple thing. I'm using a little bit of humor, but I never laugh when I say it. Or if somebody tells me that it will take four months to get something done, I might say, 'Gee, should I bring my grandmother in to help?' And whenever Grandma comes up, people know exactly what I'm talking about. I'm basically telling you you're a wimp, without saying those exact words, because I don't think that's very motivational.

"If I say, 'You're a wimp,' does that motivate you? Does it motivate you that I say my grandmother could do it? Not really, but it's just a bank shot. And believe me, I've had managers who didn't use bank shots. The cool thing is that I can be in meetings now, and I'm about to say something and somebody else will chime in and say: 'Wait, wait, wait. You know what? We'd better get Tim's grandma.' And I don't even have to say anything. All of a sudden the schedule goes from four months to three months, to two months. I think that's a really important thing, just humanizing the leadership."

Dealing with Tattlers

Here's a problem that many managers face: An employee asks to talk to you in your office. He shuts the door, and then starts complaining about a colleague who is guilty of some behavior, like not pulling his weight or undermining the project because he is wedded to a different strategy.

At that point, as a manager, you have a couple of options. You can decide to ignore the person and hope the problem goes away. (It probably won't.) You can add this thorny dispute to your already lengthy to-do list, and start down a treacherous path of playing detective, talking to people on each side to figure out, to the extent possible, who's right and who's wrong. This will chew up a lot of time, further anger people, and deepen the divide between accuser and accused. No doubt the truth is somewhere in between, and the dispute could reflect a deeper problem, while the overt disagreement is just a new flashpoint.

The good news is that there is a solution to this common problem: force people to talk to each other, and make them have an adult conversation.

It's a simple strategy used by many leaders, including Mark Fuller, the CEO of WET Design, which builds and installs elaborate fountain installations, including the fountains at the Bellagio hotel in Las Vegas.

"Early on," he said, "I decided that whenever somebody comes into my office and starts blaming something on another department, I will say, 'Really? Let's get them in here. Hold that thought.' It's just like with your children at home—you don't want serial tattling. You get everybody together, and then suddenly people are saying that maybe they exaggerated a bit, and things weren't quite as bad as they said.

"I've been in environments where a CEO will sit back and try to watch a gladiator match for entertainment. That's totally not cool. It's so common, I think, in corporate life. You want to have the conversation and say, 'Okay, what really went wrong here? There are three of us in this room. We're going to fix this thing. How do we do it?'"

Laura Yecies, the former CEO of SugarSync, an online storage service, tackles such disagreements in a slightly different way to reduce tension.

"One thing I've learned in terms of managing conflict is to challenge people about their assumptions of what's causing a conflict," she said. "Here's an example. An employee will come up to me: 'So-and-so is really giving me a hard time. They're not doing XYZ. I'm really mad.' They go on and on about how bad the other person is. It happens a million times to all managers. It's almost like what you hear from kids.

"So what I'll say to them, and it's very disarming, is 'So they're doing this thing that you don't like. Do you think they're doing it because they think that's the better way to get the job done, or do you think that they're acting that way because they don't want us to be successful?' The person will always say, 'Well, no. They probably don't want us not to be successful.' So then I'll say, 'Okay. Well, then why do they think that plan is the right thing to do? And why don't you discuss that with them?'

"I've done that, or some variation of it, a hundred times, because most people forget that they really do think that the other person is trying to do the right thing, and it's a disagreement on tactics or strategy."

Julie Greenwald, chief operating officer of the Atlantic Records Group, tells her staff that she has zero tolerance for such discussions.

"No stabbing each other. Anybody who tried that with me found out right away that I wasn't going for it," she said. "If you disagreed with someone and you were afraid to tell her, but you told me, I'd bring both of you into my office and make you talk about it. I did enough of that in the beginning that people understood that if they didn't confront another person to their face, I will blow up. If you've got an issue, you should be able to work it out. And when you can't find resolution, I'm here to break the tie, no question. But don't come to me to talk about someone else. That's not going down tonight."

The dynamic can play out in e-mail, too, with people sending notes to the boss griping about a colleague. But a firm response can shut down this behavior quickly.

"When I first got to Calvert, there was a lot of that," said Barbara J. Krumsiek, the CEO of Calvert Investments. "And one of my direct reports sent me an e-mail, complaining about something somebody else said. I just got back to them and said, 'I'm not going to read this because I don't see the person you're talking about cc'd on it. So if you cc them on it and send it back to me, I will deal with it.' Well, I never got it back, because once the person really dealt with it, it was fine."

The point, ultimately, is to establish rules to minimize destructive political behavior so that any conflicts that people have are over the tasks at hand, not with each other—a point captured by Mike Sheehan of the Hill Holliday ad agency.

"You don't want a conflict-free zone, but you want the conflicts to be about the work itself," he said. "Sometimes you have to dig a little bit and talk to people, but if you find out the conflict is about the work, then that's good, because it's healthy. I think that in a lot of workplaces it's the opposite—people have to come to a consensus on the work, and so all the conflicts are political.

"That's one thing that the founder [of Hill Holliday], Jack

Connors, instilled in the culture. It's not a democracy. You've got to make tough decisions and then you've got to move on. 'The enemy's out there,' he would say. 'The enemy's not in these four walls.'"

This blindingly simple approach to managing and leading can be difficult to implement, because it involves some uncomfortable discussions and tense moments. But in terms of saving time, avoiding headaches, and disarming destructive energy in a corporate culture, there is perhaps no more effective tool than having an adult conversation that begins with five simple words: "Can we talk about this?"

7.

THE HAZARDS OF E-MAIL

Simple conversations around tasks and team-work get lost in translation if you're not speaking to the person, and you're just texting them or e-mailing.

—Steve Stoute,
CEO of Translation LLC
and chairman of Carol's Daughter

Over the course of more than two hundred interviews with CEOs and other leaders, some predictable patterns have emerged. For example, if a CEO tells me that his company has codified its values and that there are, say, seven such values, there is a good chance that he will have trouble remembering one or two of them. (One CEO had to look them up on his iPhone.) Such moments are a powerful reminder of the benefit of shorter lists.

Here's another predictable pattern: Bring up the subject of e-mail in the workplace, and suddenly the mood shifts. E-mail is a hot-button issue, and clearly a source of endless frustration for the CEOs.

The problem, of course, is that for all the obvious benefits of e-mail in speeding up communication, it is also a dangerous trap.

E-mails are too easily misinterpreted, with often disastrous conse-
quences for the culture of an organization.

One reason that e-mail furrows the brows of so many leaders
is that the dynamic is hardly a mystery. Many commentators have
written about the hazards of e-mail, including David Shipley and
Will Schwalbe in their book, *Send: Why People Email So Badly and
How to Do It Better*. Employees should know better. And yet they
make the same mistakes time and again.

Has this ever happened to you? You send somebody what you
assume is a simple and straightforward e-mail, but your colleague's
response makes you wonder what he was reading. His back is up,
and his aggressive tone suggests that he has completely misinterpreted your intent. You scratch your head, reread the e-mail you
sent, and determine that you were, in fact, perfectly clear. Then you
start getting annoyed with the person you e-mailed for making
such misguided assumptions about how you operate. So you crack
your knuckles, roll up your sleeves, and start typing a now-wait-
a-minute rebuttal that is only going to escalate the argument.

The "cc" function can add to the trouble. Somebody raises an
issue over e-mail, others are copied on the discussion, and suddenly
everyone is picking sides like the Sharks and the Jets in *West Side
Story*.

To be sure, e-mail has its place. It is great for simple, transac-
tional messages. "When are you free to meet this afternoon?" "I'm
setting up my day, and just wondering when you'll be sending me
that report." "Want to grab lunch on Friday?"

The problem starts when something is at stake, like a person's
ego, or when there's even the slightest possibility that the intent or
tone can be misread. When an e-mail that carries a little more
weight lands in our in-box, we become amateur archaeologists,
studying the words on the screen as if we are dusting some artifact
for clues to understand what the person really meant. Why did he

copy those two people—what angle is he working? This e-mail is more terse than the one she sent yesterday—is something up? Without a clear understanding of tone, we naturally project all sorts of motives and meanings onto e-mails.

The problem with e-mail is that it does nothing to build the connective links among people that foster a sense of teamwork, and you need teamwork to innovate. And e-mail is not only unhelpful in building an effective culture, it can be very damaging. Yet the allure of e-mail is powerful, and people fall repeatedly into the same trap, thinking that e-mail is the best way to accomplish a lot of work in a short time.

Many CEOs are perceptive observers of the hazards of e-mail, and they establish a variety of rules in their companies to discourage the use of e-mail and encourage people to talk instead.

Here's what Steve Stoute of Translation LLC and Carol's Daughter tells his staff:

"I let everybody understand that they have a direct responsibility for the person next to them, and that it's very important that we're transparent with one another, that we work with one another, that we are aligned and clear in our communication. I force people to speak to the people next to them, not through e-mail. E-mail has created a lot more productivity but also has created a lot more miscommunications within organizations, because the tone is lost.

"We have to communicate, and we have to get off e-mail and pick up the phone, call our clients, and walk down the hall and speak to our peers, because tone makes a gigantic difference in the way somebody receives information. It defines urgency. It defines intent. You need tone and mannerisms to build relationships. But if you mute all those things, you start to get people who are not necessarily aligned because they don't get to know each other. They know each other by name, but they don't know each other. Simple conversations around tasks and teamwork get lost in translation if you're not speaking to the person, and you're just texting them or e-mailing."

Nancy Aossey, the CEO of the humanitarian group International Medical Corps, gives similar guidance to her employees, particularly when e-mail conversations heat up.

"I have a lot of strong opinions about e-mail," she said. "It is an extraordinary communication tool and an extraordinary way to share information. No question about it. We know what it does, so let's talk about what it doesn't do. If there's a conflict and you need to resolve it, you cannot really do it in an e-mail because people don't know tone. They don't know expression. Even if they like you and they know you, they might not know if you were irritated or joking in an e-mail. There are things we can say in conversation that you can't say in e-mail because people don't know tone and expression.

"People change when they talk in person about a problem, not because they chicken out, but because they have the benefit of seeing the person, seeing their reaction, and getting a sense of the person. But arguing over e-mail is about having the last word. It plays into something very dangerous in human behavior. You want to have the last word, and nothing brings that out more than e-mail because you can sit there and hit Send, and then it just kind of ratchets up and you don't have the benefit of knowing the tone.

"And the 'cc' function? That's the most dangerous of all. If you copy five people on your team and five people on mine, suddenly we're lining up against and around individuals. I think from a leadership position, you have to set the tone and say this is not acceptable. We're not going to conduct ourselves this way. We take a hard line that if you have an issue to resolve, you don't do it over e-mail. Do I have people who follow it all the time? Of course not. But when I see it, I flag it."

Ori Hadomi, the CEO of the medical technology company Mazor Robotics, established a rule to keep employees, particularly those in different offices, from arguing over e-mail.

"E-mail is a very dangerous way of communicating," he said. "It has a lot of benefits and a lot of advantages, but I find it dangerous because in cases of conflict, of disagreement, when people begin to argue over e-mail, they escalate their commitment to their opinion.

"So the rule we established is that I can write you something, and you may disagree, and so you might write back, 'I strongly disagree for these reasons.' Then I might write you back, and say, 'I also disagree with you for these reasons.' After that second response, there is no more e-mail correspondence. Then you pick up the phone. And we also try to use Skype videoconferences for that. It takes ninety percent less time to resolve conflicts when we talk, compared with when we write.

"Disagreeing in e-mail is just not constructive. It's not the right way to communicate if you want to come to an agreement. So that's one rule that I insist on: Let's just talk, or let's pick up the phone. And I try as much as possible to use video. I want to look at the eyes of the other person, especially when we disagree."

Another smart rule about e-mail: tempting as it may be to try to save time, do not read e-mail while you're talking to somebody on the phone, said Michael Mathieu, the former CEO of YuMe, an online video advertising firm.

"When you have a conversation with somebody, you're not going to get the nuances of the conversation if you're doing too many things," he said. "I try telling people, if somebody picks up the phone, stop your e-mail, stop what you're doing, listen and have that conversation with the person, and then move on. With most people in business, they're on the phone and they're on e-mail, and you know when they're on e-mail."

To everyone who feels overwhelmed by e-mails, and therefore lacks the time to pick up the phone, Jeff Weiner of LinkedIn has some simple advice:

"Like any other tool, e-mail is what you make it," he said. "One thing I realized is that if you want to reduce the amount of e-mail in your in-box, it's actually very simple: you need to send fewer e-mails. I know it's kind of a self-evident truth. Because every time you send an e-mail, what's going to happen? It's going to trigger a response, and then you're going to have to respond to that response, and then they're going to add some people on the 'cc' line, and then those people are going to respond. You have to respond to those people, and someone's going to misinterpret something. That's going to start a telephone game, and then you're going to have to clarify that stuff. Then you have someone in a time zone who didn't get the clarification, so you're going to have to clarify that clarification.

"So I try to clearly identify who's in the 'To' line and who's in the 'cc' line. I'm going to be as precise as possible with every word I write. I'm going to try to convey the right information to the right person at the right time. And if you can hold to that, it can be an amazing tool."

Dinesh C. Paliwal, the CEO of the audio and entertainment equipment company Harman International Industries, said that his efforts to reduce destructive politics at his company include encouraging people to talk in person rather than over e-mail.

"We're a big matrix company, and there is a lot of room for politics because matrix means you might have two bosses," he said. "If there's a problem, a conflict, I'll often go to the source. I don't talk over the phone. And maybe over dinner I'll do some coaching, talk straight, give them some advice on how to work well with the other individual, and encourage them to talk to each other. I'll tell them, 'Give him a call, talk to him more, and don't try to solve the difficult issues through e-mails, because e-mails can cause serious misunderstandings or even disasters sometimes.' And when people talk, they generally sort out things."

Phil Libin of Evernote has sought to reduce the use of e-mail at his company, to foster better communication.

"One of the things I've tried to do is uproot any sort of e-mail culture at Evernote," he said. "We strongly discourage lengthy e-mail threads with everyone weighing in. It's just not good for that. Plus, it's dangerous, because it's way too easy to misread the tone of something. If you want to talk to somebody and you're a couple floors apart, I kind of want you to get up and go talk to them."

It's simple but powerful advice. By talking over the phone or in person, you'll not only avoid dangerous misunderstandings, but you'll also develop relationships and a sense of trust with colleagues—an essential ingredient in building an effective culture. After all, it's hard to create an innovative company when the employees spend their days huddled in cubicles and offices behind computer monitors, said Tim Bucher of TastingRoom.com.

"If you see people talking to one another in the hallways, and yelling over the cubicles, and going inside each other's offices, that's a good sign. There's a certain buzz," he said. "But I've been acquired by some companies where I'd walk the hallways and it's just like a mortuary. You've got to see the interaction going on between everybody in the company, not just between the executives. And not just in meetings. How are people working together? Are they really working together as a team? Obviously there's a lot of e-mail and digital communication that occurs, and that's all good. But there's a much higher bandwidth when people sit and work together. We might sketch out something together, and we're yelling at each other, and all that stuff. And that's when you really can tell whether the culture is innovative, too."

Bucher's question—"Are they really working together as a team?"—may sound simple. But to be able to answer it in the affirmative, leaders must work through the foundational aspects of an effective corporate culture: creating a simple plan; establishing rules of the road; treating people with respect but also holding them accountable; and encouraging people to air their differences with one another (but never over e-mail). With that framework firmly in place, leaders can then focus on fostering and building a culture of innovation.

PART TWO

TAKING LEADERSHIP TO THE NEXT LEVEL

8.

PLAY IT AGAIN AND AGAIN

Never give people a void. Just don't, because
instinctively they'll think something is awry.

—Geoff Vuleta,
CEO of Fahrenheit 212

The career path to the corner office is often filled with surprising
turns and unusual beginnings. The résumés of CEOs may include
early work as English teachers, laboratory scientists, and film producers. Many CEOs have a background in theater, too, which they
call an excellent training ground for leadership.

"You need to be able to get up and deliver the good news and
the bad news," said Caryl M. Stern of the U.S. Fund for UNICEF.
"It's just that same feeling before you go onstage, and you take that
deep breath."

There's another parallel with the theater: leaders have to grow
comfortable repeating the same lines to an audience. As many CEOs
will attest, there is no such thing as overcommunicating important
themes and messages in an organization, or keeping people too
informed of the company's progress. Many leaders said they had
grown to understand that they had to make time in their schedules
for repetition and constant communication, which they consider

essential tools for creating a corporate culture that will help their organizations perform at a consistently high level.

"You have to be careful as a leader, particularly of a big organization," said Christopher J. Nassetta, the CEO of Hilton Worldwide. "You can find yourself communicating the same thing so many times that you get tired of hearing it. And so you might alter how you say it, or shorthand it, because you have literally said it so many times that you think nobody else on earth could want to hear this. But you can't stop. In my case, there are three hundred thousand people who need to hear it, and I can't say it enough. So what might sound mundane and like old news to me isn't for a lot of other people. That is an important lesson I learned as I worked in bigger organizations."

Leaders with backgrounds in academia say that one of the harder lessons they've learned about running an organization is that they must communicate constantly. Here's Amy Gutmann, the president of the University of Pennsylvania:

"I've learned over time, in an organization that's as large as Penn, with thirty-one thousand employees, that you have to say a lot and to a lot of people. I think I'm much more communicative than I used to be. I used to think that if you said something once, it should be enough. If you're a scholar and you write something once, you don't want to repeat it over and over again. There's no such thing as plagiarizing yourself, but there is such a thing as being boring. So I communicate more, and I also do more outreach to multiple constituencies.

"And I really believe in communicating in every possible way, walking around the university, e-mailing, calling, visiting people especially in their offices where they work. And I've started just randomly going to offices I'd never see otherwise and thanking people when they've done a great job but also asking them, 'What's the single thing you're most proud of?' I ask them both because

I'm genuinely interested in their answer but it's also a check as to whether our highest priorities are reflected at the grassroots level in the organization."

Marcus Ryu of Guidewire was on track to become a professor before he shifted to the business world, and he had to adjust how he communicates to simplify and repeat key themes so that employees feel fully committed to their work.

"I came from a very academic background, so I had a lot of faith in the power of words and the power of ideas," he said. "In philosophy, especially analytic philosophy, there's this idea that if you have a powerful enough argument, then you can just compel the other person to agree. That's what philosophers actually believe; that's how they interact with each other. Of course, the real world isn't like that—people have another option, which is just to ignore you, and that happens all the time. Even though we talk about how important rationality is in the company, I've come to accept that rationality plays a very limited role in persuasion, and that it's mostly about emotion. It's mostly about empathy and about authenticity and about commitment. These kinds of things are what persuade and it's the reason that people make big decisions, important decisions, like whether to bet their career on joining and staying with a company.

"Number two, sort of a corollary to that, is about communicating with large groups of people. I've come to realize that no matter how smart the people are that you're communicating to, the more of them there are, the dumber the collective gets. And so you could have a room full of Einsteins, but if there are two hundred or three hundred of them, then you still have to talk to them like they're just average people. As the audience gets bigger and bigger, the bullet-point list has to be shorter and shorter, and the messages have to be simpler and simpler.

"I was known for writing intricately crafted long e-mails.

They're still long, but the messages are generally simpler. And when I speak to the whole company, the messages are just very straight-forward and simple. One of the main things people want to hear from their leaders is optimism about the future. They want to hear truth, but when you're talking about the future, they want to hear optimism, like, 'I stand before you. I'm here for the dura-tion. We've got some challenges. We're equal to them. Nothing we have to do is harder than what we've had to do to get here. And we're going to join arms and be shoulder-to-shoulder and we're going to succeed as we never have before.' And that, in some form, is the message that you have to give every time."

Laurel J. Richie of the WNBA shared a vivid metaphor to underscore the importance of repetition.

"I keep learning time and time again about how important it is as a leader to have a clear vision and communicate it often," she said. "I'm usually very clear in my head about where I think we should be going, and I'm always learning that you cannot overcom-municate that. I get a little bored with it because it's familiar to me, but I realize it almost has to become a mantra so that everyone on the team knows where you're headed.

"You tell people, 'Here's where we're headed and these are our priorities,' and then you just sense how often people are wandering. I always say that part of the job is keeping all the bunnies in the box. You start with all the bunnies in the box and then somebody gets a great idea to go do something else and you go help them all come back and get in line and then a bunny over here pops out. So the more the bunnies are getting out of the box, the more I realize I just haven't done a good enough job communicating what our priorities are and what our focus should be."

Communicating takes time, and time is a leader's most precious commodity. The payoff can seem elusive, but many leaders under-stand its rewards. Doreen Lorenzo, president of Frog Design, an innovation consulting firm, sets aside time when she visits her offices for many one-on-one meetings.

"People want to hear from me," she said. "I thought I was being very communicative. But they want more, so I send out updates constantly. They are usually pretty passionate and it's about what's going on in the business. I've added companywide phone calls. I've added town hall meetings and then I make my rounds to all the studios. And as I do, I just try to meet people one on one. I'll set up a day and book my calendar to have one-on-ones with people, because I've found that you learn so much that way.

"So that was an adjustment I had to make. Somebody said to me, 'Well, how do you manage the time?' But you have to recalibrate. If your people are asking you for something, and you're a people-driven business, you've got to listen to that. I think that's part of being a good leader—really listening and making those adjustments and letting people know that you are serious about what you say. It's not just lip service."

I asked Lorenzo why she preferred one-on-ones to small-group meetings, which would, of course, be more efficient.

"What happens when you meet with five people or two people, there's always going to be somebody who's braver and talks more. There's always an alpha in any group, and that's okay, but you're not getting at what's on people's minds. And you learn different things. You learn about changes people want to make. You learn about their personal lives. We have a lot of people having children. So we have 'kid day' now in certain studios. People bring their kids into the studio and we have arts and crafts. These are things that we've added as a result of wanting to include families in the equation. You get that from talking to people. If I sat at my fifty-thousand-foot level, I'd die.

"In my one-on-one meetings, I'll just say, 'What do you want to talk about?' I can start a conversation pretty easily and get people talking. Oftentimes, people just want to hear about the vision. Even though you've talked about the vision five thousand times, people want to hear about the vision. They want to hear it from you. They want to be sitting close to you and they want to hear that, and that's fine. Despite all the technology we have in the world, people still need human contact. Think about it. I'm sitting with someone who is maybe newly married, maybe just had a kid, they have a house payment, and their spouse might have stopped working to take care of the child. They've got a lot of responsibility, and they're looking at me saying: 'Okay, do you get that? Is this company going to be okay?'

"I really take that seriously. I get the fact that I'm responsible for their livelihood. That's a very human thing, and I want everybody to feel confident that we're getting through that. We've come through a lot of downturns really well, and I think it's because we keep everybody very focused on doing the best they can do. If you do the best work you can do, listen to your clients and work really closely with them, and you produce something that's really great, then you're going to be successful."

Seth Besmertnik of Conductor said that one reason he communicates so much with his staff is that he treats them, in effect, as investors—specifically, as investors who are deciding where to spend their time.

"Every ninety days, we have these all-day meetings that are basically shareholder conferences," he said. "But it's only for our employees, and we go through our financial performance in detail. We go through every level of the business. My feeling is that our employees have a lot of options for how they want to spend their time. Their time is their most valuable asset, and they can choose where they want to invest that asset.

"We feel fortunate enough that they've chosen Conductor, but we also know they have other places where they can invest that time. So just like we have investors and a board, we treat employees as shareholders in a more traditional sense. We think of our employees as investors, and we reciprocate by sharing all the relevant information frequently so they understand the value of their investment, and they also understand how their contributions can grow the value of their investment. We think that if people understand why they're doing things, they'll have a sense of purpose and trust that their leaders are being open and candid and including them as equal members of the team. Then you'll have people who are more motivated and you'll have lower turnover."

Leaders don't always have to have all the answers. The simple act of getting up in front of the staff to acknowledge challenges can help quiet the rumor mill, said Victoria Ransom of Wildfire.

"As the company gets bigger, and people don't know me quite as well, I started to realize how what you say can have such an influence," she said. "You can't just say things off the cuff anymore, because people take it so much more seriously than you ever meant it. And that can be good and bad. The bad is that you might say something sort of flippant or you're trying to be really transparent and honest with the team about the challenges we may have. But that can get passed on down the line and repeated until there's a panic. On a positive note, I was surprised to learn how comforting what I say can be to the team, even if I'm not giving the answers. I thought at first that I always needed to be able to give them the solution, but I realized that actually that wasn't needed at all. All that was needed was acknowledging the challenges, and showing that we're on top of it and we get it. And I've learned the importance of addressing problems as quickly as possible, because otherwise people start to build things up in their minds, and they talk."

Therein lies one of the most important insights about the need for repetition and about overcommunicating. If employees are faced with what seems to be an empty stage, their imaginations will kick in and rumors will start up. Leaders have to take the stage and grow comfortable with repeat performances.

"If you're working in a void for any period of time, human nature says you'll view it negatively," said Geoff Vuleta of Fahrenheit 212. "You get scared; you begin to believe that what isn't there is probably bad. Never give people a void. Just don't, because instinctively they'll think something is awry."

9.

BUILDING BETTER MANAGERS

> *A boss creates fear; a leader, confidence. A boss*
> *fixes blame; a leader corrects mistakes. A boss*
> *knows all; a leader asks questions. A boss makes*
> *work drudgery; a leader makes it interesting. A*
> *boss is interested in himself or herself; a leader is*
> *interested in the group.*
>
> —RUSSELL EWING,
> British journalist (1885–1976)

In early 2009, Google launched an initiative code-named Project Oxygen with a simple goal: to build better managers. It grew out of a data-crunching effort by the human resources department, which found through studying quarterly performance reviews that there were huge swings in the ratings that employees gave their bosses. Managers were the biggest factor in employees' performance and in how they felt about their jobs.

"The starting point was that our best managers have teams that perform better, are retained better, are happier—they do everything better," said Laszlo Bock, Google's senior vice president for "people operations," which is Googlespeak for human resources.

"So the biggest controllable factor that we could see was the quality of the manager, and how they made things happen. The question we then asked was 'What if every manager was that good?' And then you start saying, 'Well, what makes them that good? And how do you do it?'"

To ask such questions was a sharp departure for the company. Particularly in its early years, Google took a simple approach to management: leave people alone, and let the engineers do their work. If they were stuck on something, the theory went, they would ask their bosses, who were in those roles because of their deep technical expertise and presumably could solve the problem. But as the company grew, the limits of the early approach became clear, and Google needed a new way to train managers and evaluate their work.

For Project Oxygen, the statisticians gathered more than ten thousand observations about managers across more than one hundred variables. They drew on performance reviews, feedback surveys, and other reports. Then they spent time coding the comments, to look for patterns. Once they had some working hypotheses, they figured out a system for interviewing managers, to gather more data. The final step was to code and synthesize all those results, more than four hundred pages of interview notes.

From that work, they drew up a list of eight behaviors of effective managers:

- Have a clear vision and strategy for the team.
- Help your employees with career development.
- Don't be a sissy: be productive and results-oriented.
- Have clear technical skills so you can advise the team.
- Be a good communicator and listen to your team.
- Express interest in team members' success and personal well-being.

- Empower your team and don't micromanage.
- Be a good coach.

At first, the list elicited little more than shrugs from Google executives. "My first reaction was 'That's it?'" Bock said. But then, the Google team began ranking those eight directives by importance, and Project Oxygen took an interesting turn.

Bock's group found that technical expertise—the ability, say, to write computer code in your sleep—ranked dead last among the eight factors. Employees most valued even-keeled bosses who made time for one-on-one meetings; who did not dictate answers but rather helped people puzzle through problems by asking them questions; and who took an interest in their employees' lives and careers.

"In the Google context, we'd always believed that to be a manager, particularly on the engineering side, you need to be as deep or deeper a technical expert than the people who work for you," Bock said. "It turns out that that's absolutely the least important thing. It's important, but pales in comparison. Much more important is just making that connection and being accessible."

Once Google had its list, the company started using it as a tool in training programs, as well as in coaching and performance review sessions with individual employees. It paid off quickly. "We were able to have a statistically significant improvement in manager quality for 75 percent of our worst-performing managers," Bock said.

The three most important habits of good managers that Google discovered—meeting regularly with employees; taking an interest in them personally; and asking questions rather than always providing answers—are echoed by many CEOs. Given the outsized impact that good and bad managers have on the performance of the people who report to them, establishing a common language and understanding around the fundamental role of managers can lift any organization.

Checking In

Part of the challenge of managing people is to find the right balance in terms of how much time to spend with them. Too much can come across as micromanaging, while too little can make managers seem aloof. Many CEOs shared stories of how they learned the value of regular check-ins.

"I found early on as a manager that it was hard to learn how to delegate," said Deborah Farrington of StarVest Partners. "I think most people in their early leadership positions either tend to delegate too little or too much. And I delegated too little at first. I felt I needed to know everything that was going on, and so I ended up doing a lot of the work myself that the people who reported to me should have been doing. I found myself working twenty-four hours a day, seven days a week. I stepped back and said, 'This is not going to work.'

"So I sat down and talked to the people who worked for me, and we agreed on various goals. But then I delegated too much. When they came back at the end of the quarter and I saw what they did, I realized that approach didn't work well, either. So I learned the importance of weekly check-ins, and then I think I got the balance right."

Many leaders will make such meetings a fixed part of the culture at their companies.

"We encourage managers to meet with all of their direct reports at least once a week to talk, to give feedback and to get feedback," said Liz Elting, the co-CEO of TransPerfect, a translation service. "These once-a-week meetings with direct reports, for each leader at every level in the organization, are very important.

"We implemented that after a few years in business, from seeing that people often had things on their minds that perhaps they would discuss with one another. They might talk about what they're not happy with, and then they'd walk out the door. And it can happen often, particularly when you have a lot of employees and everybody's busy. But we encourage managers to have these conversations because we need to keep a good, open dialogue with our people and not have them leave the company and make us wonder, 'Okay, what went wrong?'"

Cathy Choi of Bulbrite said her own experiences with a coach have taught her the importance of having similar meetings with her employees.

"I have a two-hour, one-on-one meeting with my coach each month. As I'm learning this skill of coaching, I'm passing that on to my direct reports, so once a month I meet with my direct reports for one-on-one sessions. But they're not allowed to bring paper. They're not allowed to talk about any projects they're working on. It's about their development, and where they want to go, and how I can help. It can be specific. It can be personal. It's the one hour when there's no agenda. It's their time just to talk about what they want to accomplish. And it evolves into many different things."

Angie Hicks, the cofounder and chief marketing officer of the consumer review website Angie's List, shared an important insight about the perception of feedback:

"I've realized you just have to take extra care, and make time to talk to people," she said. "When you're working with people, even if you think you've said something, maybe you need to say it two or three more times. And make sure you're praising people, and make sure that you're giving them feedback. Because their perception of how much feedback you're giving them is always less than what you think you're giving."

Questions, Not Answers

People are often elevated into management positions because they have excellent technical skills in their current job. The assumption is that they will also excel as managers. But the very quality that helped vault them into management—that they often have answers others don't—can be a handicap when they become managers. They have to learn to resist the temptation to always solve problems.

Shellye Archambeau of MetricStream said that during her regular discussions of leadership she counsels her executive team about this tendency.

"'Don't be a mama bear,'" she tells her managers. "What that means is, when people come to you with problems or challenges, don't automatically solve them. As a mama bear, you want to take care of your cubs, so you tend to be protective and insulate them against all those things. But that doesn't help. If you keep solving problems for your people, they don't learn how to actually solve problems for themselves, and it doesn't scale. Make sure that when people come in with challenges and problems, the first thing you're doing is actually putting it back to them and saying: 'What do you think we should do about it? How do you think we should approach this?'"

This approach—to ask questions, not just provide answers—is used by many CEOs to push down responsibility and to help their employees grow and develop.

"I ask this question a lot in different situations: 'What do you recommend we do?'" said Bob Brennan of Iron Mountain. "You can get a real sense for who's invested in moving the company forward, and who's watching the company go by, with that very simple question. People lay out problems all the time. If they've thought through what should be done from here, then you've got somebody who's in the game, who wants to move, and you can unlock that potential. Bystander apathy or the power of observation, in and of itself, is not very valuable. There are amazingly eloquent diagnosticians throughout the business world. They can break down a problem and say, 'Here's your problem.' But it's prescriptions that matter. So how do we move from here, and what specifically do you recommend?"

Bill Flemming of Skanska USA Building uses a similar approach.

"When people ask me a question," he said, "I don't always answer it with 'Yes, this is what I want you to do,' or 'This is what I'd do.' Instead, I'll ask them, 'So what do you want to do?' The point is that I don't want you to just announce the problem to me and expect me to solve it. You tell me what the problem is, you tell me what your proposed solution is, and I'll give you feedback. I don't always want to give you an answer on what to do. I want you to think about what your answer's going to be. Many times I've said to people, 'So you're announcing a problem? I don't want to hear what the problem is. Tell me what the solution is.' I'll always have an opinion about something, but I want people to form their own opinions.

"I've seen organizations where the boss makes all the decisions. That's not leadership, that's a boss. I don't want to be the boss, I want to be the leader. So I want to get you to help me figure out what we've got to do here. Because if you're deeply immersed in the problem or the issue, you probably know a lot more about it than I'm going to know. So what do you think is going to work? I can give you some insights based on my experience, and I can give you a different view, but you may be more intimately involved with the issue than I am. So that's another technique I use. It's captured in a quote I once read about leadership from Russell Ewing, a British journalist, who said: 'A boss creates fear; a leader, confidence. A boss fixes blame; a leader corrects mistakes. A boss knows all; a leader asks questions. A boss makes work drudgery; a leader makes it interesting. A boss is interested in himself or herself; a leader is interested in the group.'"

The questioning approach is also effective for providing feedback and for understanding circumstances that may have led to a problem. Through questioning, managers can make employees feel less defensive, said Niki Leondakis, the president and chief operating officer of Kimpton Hotels and Restaurants.

"Rather than sitting down with someone and telling them what's wrong or what needs to be addressed or what needs to be fixed, I ask how this came about and what's happening here, and listen to the back story. How'd we get here? Why does this look like this?" she said. "And then, when I have understanding about it, I can then turn it into a coaching moment rather than a moment of judgment and fear and intimidation for the person on the other side of the table who's listening to what's wrong with their performance.

"When I was a younger manager, my anxiety about what wasn't right drove me to confront things quickly. The faster I confronted it, the more quickly it could get fixed. It would get changed, and while that's a good thing, the manner in which I did it frequently left people feeling defensive. So by listening first and trying to understand how we got here and their story, I think it allows them to then hear my point of view. And then we can move into solutions. When people feel judged right out of the gate, it's hard for them to open up and listen and improve."

Seeking Out Passions

When managers interact with their employees, the meetings can easily become simple transactions: "When will that report be ready?" "Have we heard back from the client?" "What's the status of that project?" But it's important to show people you care about them, many leaders say, not just about the work they're expected to deliver.

David C. Novak, the CEO of Yum Brands, said that caring about people is the cornerstone of his leadership approach.

"What I think a great leader does, a great coach does, is understand what kind of talent you have and then you help people leverage that talent so that they can achieve what they never thought they were capable of," he said. "The only way you can do that is to care about the people who work for you. No one's going to care about you unless you care about them. But if you care about someone, genuinely, then they're going to care about you because you're making a commitment and an investment in them.

"You show you care by really taking an active interest in the people working for you, and you care enough to give them direct feedback. People are starved for direct feedback. People want to hear how they can do better. Too many leaders don't provide that feedback. So if you take an active interest in someone, you take an active interest in sharing with them your perspective on what they can do to improve."

Will Wright, the videogame designer behind such hits as *The Sims*, has an interesting take on the art of discovering each person's "special sauce" to bring out their passions.

"A lot of the people I've managed—artists, programmers, producers—they don't want to know just if they are doing a good job or not. They want to be pushed and challenged in their career," he said. "So, if they feel like you are presenting things to them in such a way that, a year later, they are definitely going to be a better artist or a better programmer, then it really feels like a win-win. Even if you give them tough critical feedback, they see the benefit and value of it, as opposed to just a typical performance review.

"For a lot of people, their job and their position are not the relevant part of how they see themselves. They have an internal view of themselves, their career aspirations, the direction they want to go. The really important motivational stuff is more in their secret identity. You get a sense of that by talking to them. You want to spend a fair amount of time exploring their interests, what they do outside of work. Usually people have some passion that really drives them.

"And this to me is one of the important points of working collaboratively with other people—trying to get a sense of what is the one thing that makes their eyes light up, they get excited about and they won't stop talking about. And if you can get a sense of what that is from somebody, and you can harness that, that's going to have more impact on how they perform their job, how they relate to you, how you can convey a vision to them in a way that excites them. For me that's the real key to a lot of this stuff—exploring and understanding the personal passions that people working with you have."

Jarrod Moses, the CEO of United Entertainment Group, says that when he takes a personal interest in his employees, he knows it will increase their commitment to their work.

"I take it personally when someone isn't happy or excited in their job," he said. "If they're not, I want to know why. I will pull somebody aside and talk to them about what's going on in their life. People have to know I'm there to support them no matter what they need, personally or professionally. A lot of my time is devoted to those conversations. But what I hope they get out of it is that they feel that they want to go the extra mile. You have to be among the players all the time. The captain of the team doesn't have to be the star. He can be the sixth man on the bench, but he has to be the one that excites people and gets them enthused about showing up every day."

Sharing a "User Manual"

Besides these three keys to effective management—asking questions, having regular meetings, and taking an interest in the lives of employees—a fourth emerged from my interviews with CEOs.

Call it the user-manual approach to management.

Computers and other devices typically come with a user guide that covers various functions, offers some tips, and provides troubleshooting advice. Wouldn't organizations operate more smoothly if everybody had a user guide that clearly spelled out his or her work style, so that people could be up front about what they want and need from colleagues and about their likes and dislikes? As it is, we generally have to figure those things out over time, almost like a detective. On the basis of experience, we piece together profiles of one another—about how much feedback people need, how they react under stress, what makes them tick.

Many leaders have explicit discussions with their employees to better understand their work styles, and to communicate what the leaders feel is important to them, so they can work together more effectively.

Tracey Matura said that she applies this approach in her work as general manager for the Smart Car division of Mercedes-Benz USA.

"I've learned the importance of building the right team, and I've also learned I have to open up about who I am, and understand how people on my team like to work," she said. "Some people need you to talk to them for the first five minutes of their day about what they did over the weekend, and you can't undervalue how important that is. I make it a point now, which I probably didn't do early on as a leader, to know what everybody's about. Mary needs me to have the five-minute conversation, and Joe needs me to just let him run because he's more like me—he's been up since six a.m. and cleared out all his e-mails so that he can hit the ground running when he gets to the office.

"I think I've developed that skill to let them know me, and for me to get to know them better. It doesn't seem like you should have to be taught that or learn that, but we're all human. So if I'm the kind of person who just hits the ground running, it's not because I mean to be exclusionary or rude; it's just that that's how I am. But as the leader, I need to understand how everybody else likes to work. So I think knowing your team and letting your team get to know you is really important.

"For me, I've had to explain that if I'm running a thousand miles an hour, it doesn't mean everybody has to run a thousand miles an hour. It's just who I am, and people need to feel free to say to me, 'Hey, can you slow down because I go a hundred miles an hour and you're driving me crazy?' And I've had those conversations. People have said: 'Remember when you told me I could tell you things? I'm telling you now you need to slow down.' At Smart, I built the team from the beginning, and I had that conversation with each new person who came on the team, and I reiterate it so that everybody understands me. I think sometimes leaders don't explain themselves, and we don't necessarily know that we should.

"I also tell them I have to stop and remember that some people want to have that five-minute conversation in the morning, and if I forget to do it, it's not because I don't care what you did on the weekend, and it's not because I don't want you to ask me what I did. I'm not being impersonal or secretive; it's just that we all go back to the way we are. And if I'm a quiet observer, and if I tell them that, they'll say, 'Oh, I know a little bit more about her.' So the things I basically tell people are for them to remind me if they need something from me. I want them to understand where it comes from. It's just who I am. I am naturally shy, which everybody laughs at, but I want you to respect who I am, and I will respect who you are and you can call me out on it because I'm your leader."

Some leaders explicitly tell their staff how to approach them, so as to remove worries among employees that they might be bothering the boss. Laurel Richie of the WNBA has developed a "rule of three" for her staff about issues they bring to her.

"I tell people that I know I can be a bit controlling, so if people are feeling like they don't have room and space, they should tell me that," she said. "I tell people I really like to be challenged. I am rarely without a point of view, and so I always tell people to come back three times if they really believe in their point of view. The first time they'll say something, I might say, 'It's a really good point but here's why we're doing this.' And if they really feel strongly, they'll come back again and I'll say, 'I really hear you so I want you to know I'm genuinely listening.' And then if they still feel that strongly, they can come back a third time and say, 'Remember when you said come back the third time? This is the third time.'

"And I tell people, don't let the conviction of my thinking ever get in the way of feeling like you can challenge me. I have learned that I speak very forcefully and very directly, and for a lot of people, that says to them, 'Okay, that's done. That's off the table.' And I really don't mean that it's off the table. I get it from my dad. I speak directly and it sounds like it has been decreed as opposed to that's my thinking today and that's my point of view.

"I developed the rule about coming to me three times because I was hearing too many people saying, 'Yeah, we really felt strongly about it but you made it pretty clear that wasn't an option.' So that made me realize there's something in how I'm acting that people don't feel like they can challenge me even though I'm saying I want to be challenged."

Chauncey Mayfield of MayfieldGentry Realty Advisors said that he learned early on to be comfortable asking people questions, and that he applies this approach to better understand how his employees like to work.

"As a kid, I spent time working in my dad's law firm—he was a criminal lawyer—as a gofer," he said. "Oftentimes, when clients would come in and if I didn't have anything to do, I would sit in the lobby and talk with them. I was curious. I'd ask them: 'What did you do?' 'Why are you here?' I became very comfortable over the years with older people. I always thought I could learn something from them. 'Tell me, teach me—how does this work?' I have the same philosophy today. I have no idea, for example, what irritates my administrative assistant. I'll say, 'You have to tell me, specifically, what makes this work for you and what doesn't make it work for you. I'm not going to try and figure it out. I know what I need, but you've got to share with me what you need as well.'"

At Continuum, an innovation design consulting firm, CEO Harry West said the company has long used a structured system to help employees better understand their colleagues' work styles.

"One thing we have done for about a decade is social-styles training," he said. "There are some very simple questionnaires that elicit insights from people about their communication style. We ask everybody to go through that, and then we share that information with everybody. We even put it on our intranet because we want people to understand that often people don't necessarily disagree with you, it's just that they have a different way of expressing the idea. So we try to tease apart the issues around communication style from the issues around content, because we want to focus on the content, not necessarily the style.

"We have an incredibly broad range of people in our company, from people in their twenties to people in their sixties, and they have backgrounds in design, anthropology, electrical engineering, software, graphic design, law, history. This can mean differences in communication style. It's just the way it is. So we need to get over that and focus on the content. You might learn that one person is always going to phrase every question as a question, and another person is going to phrase every question as an option. And they're both trying to say the same thing, it's just that they have different ways of doing it."

After hearing of this approach from many leaders, I interviewed Ivar Kroghrud of QuestBack. He took the notion of a "user manual" one step further. He actually wrote one. I'll reprint it here, and then let Kroghrud explain his thinking.

How to Work With Me
A "user's manual" to Ivar Kroghrud

For best results:

- I am patient, even-tempered, and easygoing. I appreciate straight, direct communication. Say what you are thinking and say it without wrapping your message.
- I am goal-oriented, but have a high tolerance for diversity and openness to different viewpoints. So again, say what you are thinking and don't be afraid to challenge the status quo.
- I welcome ideas at any time, but I appreciate that you have real ownership of your idea and that you have thought it through in terms of total business impact.
- I have a very high trust in others. I am accepting and trusting and always want to think the best about people. As a trusted team member, be aware of this quality and remind me if you think I'm being led down the wrong path.

- I tend to be fond of deliberation and reflection—i.e., I am cautious, thorough, and sometimes seem slow in decision making. This is often good and ensures quality decisions, but given the points above: I expect you to kick me in the ass when you feel it is called for.
- I tend to shy away from conflict and confrontation. I sometimes accommodate easily to the needs of others when challenged. I am aware of this and I am working on it. As a trusted team member, please help me.
- I am very much a team player and work best when in team settings. Achieving things together with others is what makes me tick.

How to use these points:

Be aware of them and use them in your dealings with me. The points are not an exhaustive list, but should save you some time figuring out how I work and behave. Please make me aware of additional points you think I should put on a revised version of this "user's manual."

Warm regards,

Ivar

Kroghrud explained what motivated him to write the document: "Writing one made sense to me because I've always been struck by this sort of strange approach that people take, where they try the same approach with everybody they work with. But if you lead people for a while, you realize that it's striking how different people are—if you use the exact same approach with two different people, you can get very different outcomes.

"So I tried to think of a way to shorten the learning curve when you build new teams and when you bring new people on board. The worst way of doing it—which is, regrettably, the normal way of doing it—is that people just go into a new team and start working on the task at hand, and then they spend so much time battling different personalities without really being aware of it. Instead, you should stop and get to know people before you move forward.

"Some of this comes from the experience I had in the Norwegian navy. Part of the leadership training was really about getting to know yourself and how you react under different circumstances and learning how different people really are. So if you understand yourself, you can start learning more about your team. So my question was 'How can I shorten this cycle? What simple things can I do to make these people work more effectively with me?'

"If you give them this 'how to work with me' page on the first day, they get a different perspective on who you are and how you relate to people and how open you are. And it's much easier when they understand that these are the things I like, these are the things I don't like, and this is how I am. The reaction from people has been a hundred percent positive. I think it just makes them open up. And there's no point in not opening up, since you will get to know people over time anyway. That's a given, so why not try to be up front and avoid a lot of the conflict? The typical way of working with people is that you don't share this kind of information and you run into confrontations over time before you are able to understand their personalities."

10.

SURFACING PROBLEMS

As a consultant, I learned that you really have to listen to the lowest-level people in a company, because that's where the answers are.

—KEN REES,
CEO of Think Finance

One of the greatest challenges for leaders, whose jobs are inevitably isolating, is to find out what's really going on in their companies. After all, people generally want to bring them only good news: "Two thumbs up, boss, everything's great!" How do you get employees to be candid with you about what's happening on the front lines? For leaders, this is part of the endless challenge of creating a corporate culture where feedback is encouraged in all directions. In chapter 6, CEOs discussed the importance of having "adult conversations" to give employees feedback. In this chapter, they describe their strategies for getting feedback—about how they can do a better job leading, and about any problems deep inside their organizations.

Dennis Crowley of Foursquare will ask his employees over coffee how he can do a better job.

"As the company has grown, I can sometimes start to feel disconnected, and I'll decide to randomly meet with one person a day, and we'll go out for a half-hour coffee," he said. "You do that for six weeks or so, and then all the channels of communication are open again. People feel like they can come and talk to me. I learn about the things that are troubling them or challenging them, or questions they might have. I always ask them for feedback, too. 'Is there anything that I can do better to make your job easier?' 'Is there anything I can do to make the company better?'"

Ivar Kroghrud of Questback has told his employees that he considers part of his role to be the company's "chief ironing officer," which encourages them to speak up about how he can smooth out processes so they can do their jobs better.

"It's very easy in a fast-growing company and fast-changing industry to get hung up on all the things that aren't working and that we should be fixing," he said. "If you want to get extraordinary results, you have to play to people's strengths and you have to help them work as close to plan as possible. If you allow them to get bogged down in all the problems that are out there—and there are always problems—then they'll be unproductive.

"You can get a lot of speed by thinking of yourself as a chief ironing officer. Once you have a successful system in place, you can spend some of your time just walking around talking to people and asking, 'What's preventing you from doing an even better job? What are you spending time on that you don't feel you should be spending time on?' Those kinds of questions are easy to ask and people relate to them."

Ken Rees, the CEO of Think Finance, a firm that develops financial products, instituted regular meetings called Cookies with Ken for employees at all levels.

"As a consultant, I learned that you really have to listen to the lowest-level people in a company, because that's where the answers are—with people who are interfacing with the customers. A lot of what consultants do is stuff that senior management ought to be doing, which is really communicating effectively at all levels and putting those ideas together. That's a big part of my approach as a CEO. I find that hierarchy's a killer, particularly in a fast-paced company. You have to go out of your way to keep ripping out those levels and barriers between people.

"Some of what I do, frankly, is a bit corny. I have Cookies with Ken every couple of weeks, and we bring in about a dozen employees from across the company and talk about what's going on. At the end, I always ask, 'Tell me one thing you really like about the company and one thing that frustrates you about the company.' I always come out with at least one thing that is eye-opening."

Abbe Raven, the CEO of A&E Television Networks, runs regular meetings she calls Raven's Roundtable.

"I started it when I became a CEO because I felt as though there was already an opportunity for us to talk to new employees. But what about current employees?" she said. "So I'll have breakfast or lunch or coffee with a group of people at different levels, people I would not normally break bread with. When I first became CEO and I was trying to talk to as many employees as possible, my opening question always was 'If I had been a CEO that came from the outside and you were meeting that CEO for the first time, what would the topics be that you would talk about? What should we change, and what shouldn't we change?'

"And if I see a new employee in the hall after they've been here for three months, I'll say: 'What is working for you here that you didn't have before? Is there something that you used to do at your old company that we should be doing?'"

Stephen Sadove of Saks also asks new hires to give him their fresh perspective on how the company operates.

"I think some of the best ideas come from people who aren't stuck in their ways," he said. "I always tell people new to my organization when they come in, 'I want you, in your first three or four weeks, to jot down every time you have an idea or a question about how things are done, and then stick it in your drawer.' Just, whatever it is, 'Why are they doing it this way?' I don't care whether it's good or bad; I don't want you to even talk to anyone about it.

"Just write it down and stick it in the drawer. And at the end of three or four weeks I want you to look at the sheet. Maybe you'll say: 'Now I understand that. Now it makes a little bit of sense to me.' Or you may look at it and say, 'That still doesn't make any sense to me.' Then I want you to sit with me and we're going to talk about them. Invariably, I find some really good ideas that make you say, 'Why are we doing it this way? It makes no sense at all.' I've seen little things, big things, waste in the system and a lot of duplication of work. Things like that come out of it."

John Donahoe, the CEO of eBay, uses exit interviews to get unvarnished perspectives on how the company is doing.

"One of the things I've also found really useful over time is any time a senior person leaves, or sometimes a mid-level person, I'll often reach out and say, 'Hey, would you either send me an e-mail, or I'd love to get together and I'd love to hear the three things that you think I should know about in the organization that I might not be aware of.' And then, secondly, 'If you were me, what would you do differently from what we're doing?'" he said.

"And I find that when people are leaving, they're often in a very reflective state and, because they've often made a very difficult decision, they're also just stunningly direct, because it's like they have nothing to lose. In fact, if they care about the company they'll be more direct. And I find I get some very good insights, and they're often quite actionable."

Susan Lyne, the vice chairman of the online retailer Gilt Groupe, adopted an idea for "office hours" that she learned about from

Marissa Mayer while Mayer was still at Google, before she left to become the CEO of Yahoo.

"She was a professor before she came to Google, and she kept office hours going," Lyne said of Mayer. "She said it was a really useful way to access the engineers' real ideas. Not the ideas that come out of a meeting, but 'What are you excited about?' And it sounded like an interesting concept to me. I do it now—I try to do it two hours a week, where anyone from our company can book half an hour with me.

"It's turned out to be a fantastic way to find out what's bubbling under the surface and what's not coming across to people. And a surprising number of people will book time with me who are significantly down the food chain. In some cases it's because they want to have a little face time with me so that they can get noticed. But there's always something on their mind.

"And when you are running a company it's very hard to get below a certain level, maybe one level below your direct reports. It does give me a way to get to know a little better the people I pass in the hallway or I see in the Monday all-hands meeting. It's also a great early-warning system for something that may be either misunderstood or a challenge within a department."

Ori Hadomi of Mazor Robotics has found that designating someone to play devil's advocate helps guard against groupthink in meetings.

"Before we set the objectives for the next year, we have a tradition where we define the five biggest mistakes we made last year—and we'll focus on the big ones, not the small ones. And every year we look to see if there is something common among these mistakes. Then we set the objectives for next year," he said.

"One of the most obvious mistakes we found is that too often we choose to believe in an optimistic scenario: we think too positively. Positive thinking is important to a certain extent when you want to motivate people, when you want to show them possibilities for the future. But it's very dangerous when you plan based on that. So one of our takeaways from that was to appoint one of the executive members as a devil's advocate.

"He's actually very challenging and he knows how to ask the right questions. He really makes sure to say to me, 'Let's be more humble with our assumptions.' And the most surprising thing is that he's the VP of sales for international markets. You would expect the VP of sales to be pie-in-the-sky all the time. But he has a very strong, critical way of thinking, and it is so constructive. I feel that in a way, one of the risks of leadership is thinking too positively when you plan and set expectations."

At Medline Industries, leaders established a system they call "escalate." Andy Mills said it provides top executives with a front-line perspective on problems:

"I like to tell employees that you can give an answer or you can give a solution. A lot of people might think they're doing a job by giving an answer. But I really don't believe that simply giving an answer is doing your job, because if the answer doesn't help move something forward or help the business, then what good was that answer? If we're back-ordered on a product and it's not due in for a while, then just telling somebody that is not a solution. It's just an answer. You haven't completed the job until you've helped the person move forward whatever it is they're doing.

"We also tell people that we want issues like that escalated. Within sales, the escalation policy is that if a customer or sales rep calls and they want something, then we want to try to find a way to do it. And if we can't, then we escalate it to their superior and it goes up the chain. Out of over nine thousand employees, there are just a handful of people we want to have the authority to give a negative response to a rep or a customer's request. Plus we want to know about the issues and not filter the problems out.

"I'm happy to take these calls, and I do take a lot of them. I find sometimes there's something we aren't doing but we should consider it. It might be a customer need that nobody in our industry is meeting, including us. So when possibly there's something we aren't doing, an unmet need, we'll really consider changing our policy or changing something in the company. And if a competitor is doing something we aren't, then we need to be able to do it, too. And even if we're not going to do it, I think people appreciate getting an answer from somebody at a high level about why. Or maybe it's a good idea that we're going to work on in the future. I think people appreciate an explanation about why we can't meet their request more than just getting told 'no.'"

There is an art to setting a tone that disarms employees so that they will provide candid insights and offer their opinions about how to improve the organization. Here are some other approaches that CEOs use:

"I used an executive coach and got some advice on coming into a new organization," said Barbara Krumsiek of Calvert Investments. "The advice was to ask each executive, 'Tell me about your job, but now tell me about what you think you do here that is not in that job description that you think is really critical.' Wow, did I learn a lot about them, and it was very informative in shaping the team. I also asked this a lot my first couple years at Calvert: 'Tell me one thing that's going on at Calvert that you think I don't know and that you think I should know.'"

Dan Rosensweig, the CEO of the textbook rental company Chegg, said he asks employees, "'If you had my job, other than giving yourself more vacation and a raise, what's the first thing that you would do that you don't think we're doing yet?'

"I also try to make it comfortable during the review process by asking people: 'What do you need more of from me? What do you need less of from me? What is it that I'm doing that you would like me to stop doing completely? And what is it that I'm not doing enough of that you'd like some more of?' From there, it becomes a much more comfortable conversation," he said.

Barbara DeBuono, the CEO of Orbis International, a global group that helps treat and prevent blindness, has used a clever ice-breaker to get people to open up.

"I've had this conversation with groups of people at Orbis, where I'd say, 'Do you think we're a high-performing organization?' and then I shut my mouth," she said. "And I didn't want to give the answer. I wanted them to give me the answer. I also asked them, 'What do you think a high-performing organization would look like?' That's how I opened the talk, and one of the key elements is honest, open, transparent dialogue, and we'll talk about the tips and tools for doing that.

"It's interesting. I'm seeing the energy level is different in the office. This is not a carrot-and-stick kind of approach, with me say-ing: 'We are going to be a high-performing organization. We are going to do these five things, and that's it.' The next questions I ask are: 'Do you want to be one? And if so, what is a high-performing organization? Let's discuss what it is.' But I definitely see some spring in people's step. I see a lot of people saying, 'It's a new day.' I'm not hearing much anymore, 'That's the way we always did it,' because they know I don't want to hear that. So they'll sometimes apologize for that. They'll say, 'I just want to tell you this is the way we do it, not that we have to do it going forward.' So that's a change, right? I'm seeing baby steps."

Kevin Sharer, the former CEO of the biotechnology company Amgen, asked his employees five questions when he took over to get their candid feedback.

"I posed several questions to the senior staff: 'What are the three things you'd like to make sure that we keep?' 'What three things would you like to change?' 'What is it that you would like me to do?' 'What is it you're afraid I'm going to do?' And then, finally, 'Is there anything else you want to talk about?'" he said.

"I talked to the top hundred and fifty people in the company one at a time for an hour. I invited them to bring in, if they wished, written responses so I could have them. I took very careful notes. I wasn't trying to sell anything. I was trying to listen deeply. And in a science-based company, we value data, and this was social data of profound importance. So I synthesized all the answers and I wrote, just before I became CEO, a summary: Here's what you guys told me and here's what I think about it, and so here's our priorities. People were enthusiastic and honest in talking to me, and it also helped me to get to know the top people in the company better."

Geoffrey Canada sets the tone at Harlem Children's Zone by telling people they should never hesitate to bring up a problem.

"We have eleven thousand kids," he said. "The thing that I've been trying to say to the senior team is there's no downside for bringing a problem up and asking for help and support. Or even consultation on it. You'll never get a mark against you, like, 'Oh goodness, this is the third time he brought something up and said he was having trouble.' I tell the senior folk that you've got to be much more active in actually making sure people really get this fact—that there's no penalty for bringing something up."

Shawn H. Wilson, the president of Usher's New Look Foundation, developed a shorthand to make sure people would be frank with one another.

"A few years ago, I started talking about a 'no-spin zone,'" he said. "I always tell my staff, 'When you come in my office, you're in a no-spin zone. Just be respectful.' I've seen the habit in other organizations, and I saw it creeping into our organization, where people tend to make excuses or spin the truth—'Well, this did happen, but it's because of this.' So at one point, I felt it was important as a leader to say, 'Listen, I don't know why this happened, but we need to get to the core root of why it happened, and it has to be factual. It can't be all these other things.' When we started that, I definitely saw a difference in the culture."

Pamela Fields, the CEO of Stetson, the hat and apparel company, tells employees that she will take the bullet for any mistake they might make. In return, they feel more inclined to support her, and to be candid with her about problems.

"The lesson I learned is the importance of telling the truth, and being in an environment where truth-telling is valued," she said. "It takes a lot of spine to tell the truth, especially in a large organization, where obfuscation is a political skill that I don't have. I see a problem, I see an opportunity, and I want to go for it. Business is too fast and we have to move.

"People who work for me know I have their back. I think that's the most important thing I can offer people—that if something goes wrong, it's my problem; if something goes right, it's their success. I say it to everybody. I tell them that if they do something wrong, and it's on my watch too often, then I clearly won't be able to hang around to help them. So we have to work with one another. But I say this explicitly. I am always there; I will always put myself in the line of fire for people. There's never been an exception. I take the bullet. That's my job. So I'm very loyal.

"And people know that they can come to me and let me have it if they think I'm wrong. And I love that. I love people saying, 'That is the silliest idea and here's why and here's what we should be doing instead.' If that conversation happens, it's a success. I can't always promise them I'm going to agree that what I'm advocating is so stupid, but there's a complete fear-free zone. I also tell people that I don't like surprises. Keep me in the loop. If it goes wrong, it's my problem, not yours. It really seems to free people up. It sounds trite, it sounds clichéd, but it's really not. I found that no matter what culture I've gone into, it's worked."

Martha S. Samuelson, the CEO of the Analysis Group, a consulting firm, said she sets an example for the behavior she wants from employees, so that they don't hesitate to bring up problems.

"A lot of it stems from this model of how the founders were with me," she said. "They always made me feel like it was okay to say: 'I'm concerned about something. I'm worried I didn't do it right. I shouldn't have said that in that meeting.' They were doing that themselves, and they were a role model for me. So I do that a lot, and I feel like I invite people to come to me with a problem. They never feel they're going to get in trouble if they come to me and say, 'I blew it with a client. I said the wrong thing at the meeting. I did this. I did that.'

"I think that runs up and down the organization, and that makes a huge difference. You want to create an organization where people feel it's safe to bring problems to their managers and ask for advice. You don't want it to be like you're putting a Band-Aid over a piece of glass in your arm, because the piece of glass is still there. You want to have somebody help you take the piece of glass out."

Dennis Crowley of Foursquare will occasionally open discussions with his staff by saying the company is "broken." It's a smart tactic, because it quickly acknowledges that a problem exists and shifts the focus to discussing how to fix it.

"We've taught people that it's okay to be critical," he said. "We try to air all that stuff in public at company meetings, and I think it creates a really healthy environment so that people aren't running off to a conference room and saying, 'I can't believe we're doing this.' If you want to talk about that, talk about it in public. That's one of the things that has made it easier for us to grow quickly and still feel relatively small.

"We're pretty critical and self-aware about where we are as a company. We always talk about when the company feels broken—

let's say you have ten employees, and suddenly you have five more, and the stuff that worked at ten doesn't work at fifteen. So we'll say, 'Okay, the company is broken—let's step back and figure out how to fix it,' and it might happen again from twenty to fifty, from fifty to seventy, whatever the numbers are. But we've been very transparent as an organization about how we're going to have to change some things.

"The teams might be too big. Maybe there are too many reviews. There are all these little levers that we can tweak, and that's how you take something that's feeling a little bit broken, or not as efficient as it could be, and right it. And the way we're self-aware really helps when you get to this stage because you're not embarrassed to say, 'The way we're doing this is not working right now, and we have to change some stuff.' Then it doesn't rock the boat when we mention it to the staff. It's easier because we've done it before."

David Sacks of Yammer said he actively encourages a culture of dissent at his company.

"Anybody can ask me questions and debate me," he said. "You could be a new employee and you can start getting into a debate with me about something. The start-up culture is very democratic in general. I think you need that in order to attract good people. You've really got to create a company culture that people want to work at. And so you try to give them a voice, give them a sense that they influence the direction of the company, and try to avoid unnecessary process and hierarchy—things that might frustrate employees.

"I think you've got to create a culture in which dissent is valued. And there's probably a lot of ways to set that tone. Certainly you can tell if you've got a culture of dissent when you walk into a company. People can figure out very quickly whether dissent is encouraged or whether it's actually not welcome. It's a red flag to me if there's just too much consensus and not enough dissent. I feel like in any human community there's always dissent because people just disagree. Anytime there doesn't appear to be dissent, it means that the corporate culture has just shifted way too much toward consensus.

"You've got to constantly ask your reports whether they think we're on the right track, whether the strategy you've laid out is right, what they think about the strategy, where things aren't going well. You've really got to dig into that. We let employees voice their opinions about everything. There's no sense that 'Okay, I am an engineer, therefore I can't voice my opinion about what's happening in customer service or sales,' or vice versa. We try to create clear ownership so everyone owns an area. One of the ways political cultures develop is when it's not clear enough who owns which areas, and so you need to get lots of people on board to do something. That's not true at our company. One reason people can feel comfortable about dissent here is because their own responsibilities are clear.

"We do a lot of things to try and pull the company together and make sure that we're all on the same page. So about once a quarter, I give a company presentation that lays out our thinking at a high level about the strategy. And then once a month we have Yammer Time at the end of the day on Friday, and the executive team takes questions from anyone in the company. They can also submit them online. They can also submit them anonymously if they want. We'll basically answer anything that people want. People can see the anonymous questions online, and people can vote on which questions they want us to answer."

Gregory B. Maffei, the CEO of Liberty Media, said that rewarding people who might disagree with him in meetings sends a clear signal that he wants people to speak up.

"There has been a general change in a lot of organizations," he observed. "There's more transparency, more openness, and at least some of the trappings of the imperial boss have been reduced. And I think that's good. I try very hard to do the things that I appreciate, like being direct about what the organization is doing. We make sure we have quarterly meetings that are very open and encourage questioning, so people feel like they are part of the organization.

"I've always felt most comfortable in a culture where people do feel, regardless of the size of the organization, that there is an ability to have dialogue, and that there is an ability to feel like you can ask the CEO any question. Too much formality or reverence can get in the way of a good exchange of ideas. So how do you make that happen? You've got to somewhat walk the walk and talk the talk. In our meeting with all employees, I try to be candid about what we did right and what we did wrong in the quarter, what's the longer term, how we're doing and what some goals are. And you try to get them to ask tough questions.

"So I usually make sure there's at least one or two that I know somebody will ask, that are going to be viewed as tough, because they want to make sure they get covered. Because you want to set a tone of, 'Hey, people get to ask those questions.' I know there are some people who are more comfortable asking the tough questions, and I'll say, 'Hey, do you think that people know enough about this? If you don't, maybe you want to ask about it.' You do that a couple of times and people know that somebody can ask a hard question, and you give them a straight answer as best you can.

"You've got to give credit for ideas that others have, and make sure you show people you appreciate them. One way is to say, 'Look, I thought this, but Albert said that, and he's right. I agree. Albert's got the right answer. Let's go that way.' Here's another example: At Liberty, one executive challenges me more than everyone else.

I respect his opinion enormously, but he's not always easy with his challenges. And he was the first guy I gave a promotion to. Two of the other executives came up and said to me, 'Well, if we knew beating up on you was the way to get promoted, we would have done it, too.' But they probably used words that were more graphic than 'beating up on you.'"

11.

SCHOOL NEVER ENDS

I think the best leaders are teachers.

—David Barger, CEO of JetBlue

We spend most of the first part of our lives in school, and the knowledge and skills we acquire there help make us productive and effective employees. For most of us, formal education ends as our focus shifts to the work of meeting goals and deadlines. But many companies recognize the need to keep their employees energized, with a sense that they are growing and learning new skills and building new muscles. One way to do this is to erase the dividing line between school and work.

Many leaders also understand that creating an environment of continuing education will help retain their brightest employees, who can quickly tire of the status quo and of doing the same work over and over.

"If you look at why people in general leave companies, they often leave because they get bored," said William D. Green, the former chairman and CEO of the consulting firm Accenture. "And high-performance people are learners by nature. And as long as they're learning, they'll stay where they are. When they start to think about leaving, when they start to respond to a headhunter's call, is when they haven't been learning."

Every company has to adopt a philosophy of constant learning if it hopes to succeed in today's economy, he added.

"With a motivated learner, you can work wonders. In institutions, it's the same thing. Are there companies with the will and resolve to change? That's the equivalent of a motivated learner. Or are there companies that are just sort of stuck where they are, and they like the status quo? In the end, that's the difference between winners and losers in corporate America and around the world. That's the contrast. So, the question is, How do you get motivated learners?

"One of our principles is that people who are successful are the ones who ask for help. It sounds simple, but to get an organization to believe that asking for help is a sign of strength, and not weakness, is a huge thing."

This is another challenge identified by many CEOs: how to foster a culture of motivated learners.

Move People Around

At many companies, leaders set an expectation that employees will be shifted to different positions. That way, individuals stay on their toes rather than settle into a comfort zone. The payoffs of this approach are many, according to Mark Fuller of WET Design.

"We move people around a lot into different positions," he said. "And quite honestly, it's pretty unsettling because everybody loves to be comfortable. I think we're built that way. Find your cave, and draw some nice picture of a mammoth on the wall so it feels like home. Most of my key people have held really different positions. The world is driven by change, so part of my job, I think, is to stir things up."

To stir things up at WET Design means all employees also go through training programs to get hands-on experience of what their colleagues do, and an appreciation of it. As Fuller explains:

"We do take all of our key employees and put them through an immersion program that typically lasts six weeks. I can show you some great receptionists who are pretty darn good welders because they spent a week or two in the machine shop. They get it, and they understand what's going on. Again, these are not permanent assignments for everybody, but it's really about walking in other people's shoes to understand their jobs."

Julie Greenwald of the Atlantic Records Group learned first-hand the benefits of doing many different jobs by starting on the bottom rung in the music business. So she lets her employees know that they should expect to move around, too. She recalled how she rose up through the ranks:

"I met Lyor Cohen and started at Rush Management, which is crazy because I didn't know anything about hip-hop music or rap music. I was Lyor's assistant. I didn't even have a desk and a table. I had to literally sit on the arm of a couch. He would be on the phone talking, yelling, barking, and I would be doing itineraries, tracking down artists. Whatever he asked me to do, I did it. Lyor used to give me crazy, impossible tasks to do.

"Lyor then sent me from Rush Management to Def Jam records. I went to the promotions department and I saw that it was very disorganized and chaotic, so I pulled everybody together and I got things organized. Very quickly, they promoted me to be general manager of the department. Then Lyor said to me, 'I need you to start a marketing department.' And I said, 'I don't know anything about marketing.' And he said, 'Go learn.' And then he handed me the video department, and I said, 'I don't know anything about making videos,' and he said, 'Go learn.' And then he handed me the finance and budgets and the art department, and before I knew it, he handed me every department in Def Jam.

"It really taught me that the best thing I could do for my company and my people is to move them around and have them learn what everybody else is doing, so they really understand the nuance of each person's job and how tough it is. That way, people respect each other on a whole new level. Everybody has to kind of know everybody else's job. They don't actually have to do it, but they have to have a real understanding, so there's a whole layer of respect to keep everybody working hard together."

Irwin Simon of Hain Celestial Group said he works hard to make sure people aren't pigeonholed in roles.

"Within corporate America, I learned that people are put in a box—you are an accountant or you are a marketing person, and basically, you were labeled for life," he said. "So I love to take a marketing person and put them in finance, or I'll take a finance person and put them in marketing. I believe in taking people out of their box, taking them out of their comfort zone and putting them into other areas. That's how you grow in life. Look at the exciting opportunity for you. I love people to move to other parts of the world, just to see how the world works. I try to push people to do other things and see other things. I'm a big believer in the idea that you've got to push people. If you don't push them, they get very comfortable."

Kathleen Flanagan, the CEO of Abt Associates, a consulting firm, said that younger employees today expect to move into new roles quickly, and so companies have to be careful to strike the right balance of managing their expectations while also providing the opportunities they want.

"I think the younger generation obviously wants to move a lot more quickly in positions than maybe the more senior folks like me. They're constantly curious about what they can do next. They're almost impatient about sitting in a job for any length of time, and they always wonder about the next opportunity.

"There's obviously tension there. You want somebody to commit for six or twelve months to something, but my objective is to build a culture that allows people to move when it's right, or at least to have those discussions. So the coaching has to be there. The managerial talent has to be there so that they can have that discussion with somebody and say, 'Here are the pros and cons of this,' and then allow them to move quickly, as opposed to just saying, 'You should stay in that job for three to five years.' This generation isn't going to do that. They'll go somewhere else. So keeping talent at our company is a real priority right now. It's very competitive so we've got to think about ways to keep that next generation growing, learning, excited."

Ensure Constant Training

Many companies go the extra step of setting up formal and informal training programs for employees. They feel it's their responsibility to help employees grow, beyond specific work skills.

"We think a lot about personal development," said Kris Duggan of Badgeville. "How can we invest in our employees to have them feel like they're really growing and developing while they're working with us? And so we do a couple of things. We will do management training for any manager. If they want to go to any conference or any training or any outside education, just let us know, and if we can pay for it, we'll pay for it. We try to have guest speakers coming in on a monthly basis to talk to the team. I think those are the things that people value, where they have a sense of growth and development. I think that's what really motivates people.

So if they feel like we truly, genuinely care about their development, I think that creates very strong retention."

Kathleen Flanagan of Abt Associates encourages employees to seek outside perspectives so that they keep learning. Her philosophy is based in large part on her early leadership lessons.

"My predecessor as CEO, who was the executive vice president who put me in that job in 1989, was a huge mentor to me," she recalled, "and he was the one who said, 'Plan for success. Make your plan, set your goals, but plan for success. Be flexible along the way, but have a game plan and make sure you're planning for success. Don't plan to fail.'

"And always strive to have butterflies in your stomach, he said, because then you'll be learning. You'll be pushing yourself. Listen and get input from a lot of people. Respect everybody for their contributions. He helped develop me as a leader and gave me the right level of support, but also put me out there to stretch, to grow, to make my own decisions. He never tried to make decisions for me, and I think that's really important.

"He surrounded himself with a team that he had confidence in, and he allowed the team to do its job; I think that's essential. I've built my team over the last couple of years. I'm very lucky because we've got a terrific team, and they are much smarter than me on many things. They're experts in what they do. And [my mentor] taught me humility. I don't have to know it all. You've got to ask for perspectives. Go with your gut at the end of the day if you have to make a decision, but get advice from a wide range of people and value their input.

"I tell people to look for ways to develop and grow. Get outside perspectives. That's a big thing that I've been pushing the last couple of years. I have asked everybody to look outside the company, to go talk to their counterparts at other organizations. Just get out there. Learn different perspectives. Go talk. It's amazing how welcoming other CEOs are to me and how willing they are to share, even if I'm viewed as a competitor."

Phil Libin of Evernote formalized a training process at his software firm after he learned about a military program.

"We recently implemented something called Evernote Officer Training," he said. "I got this idea from a friend who served on a Trident nuclear submarine. He said that in order to be an officer on one of these subs, you have to know how to do everyone else's job. Those skills are repeatedly trained and taught. And I remember thinking, 'That's really cool.'

"So we implemented officer training at Evernote. The program is voluntary. If you sign up, we will randomly assign you to any other meeting. So pretty much anytime I have a meeting with anyone, or anyone else has a meeting with anyone, very often there is somebody else in there from a totally different department who's in officer training. They're there to absorb what we're talking about. They're not just spectators. They ask questions; they talk. My assistant runs it, and she won't schedule any individual for more than two extra meetings a week. We don't want this consuming too much of anybody's time."

Marjorie Kaplan, of the Animal Planet and Science cable networks, said that teaching her employees new skills is part of her leadership philosophy of bringing out people's "best selves."

"We want to make sure we're a learning organization," she said. "We have a lot of junior people, and I noticed in some meetings that these people were smart, but that they really needed some presentation skills. I think it's great that we're in an organization that allows all these junior people to stand up and speak. But let's make it interesting, and let's let them learn.

"We put together a presentation-skills class, but we did it with storytelling. So we brought in people who specialize in storytelling. And they focused on how to tell stories and how to use those stories in presentations. That's for a couple of reasons. One is that we are a storytelling business, so I want us to be thinking about stories. A second is that I think storytelling gets you closer to yourself. I think the best presenters are people who are themselves. It gave them some skills. And so when these people are now in meetings, they're different in meetings. Their voices are stronger. They're braver."

Robert J. Murray, the CEO of iProspect, a digital marketing company, said his company has set up different forums where employees can learn, particularly from one another. They routinely hold "shared information classes." He also has held management roundtables to have employees share stories and learn from each other about how to manage people.

"I would get the young managers together in a room every two weeks and we would share examples of times they had to deal with a difficult employee situation," he said. "I felt it would resonate more with them if they saw each other learn how to manage people. So I would go around the room and say, 'Someone give me an example where they're looking for advice. Let's open up. Let's put an example on the table and let's all talk about it.' And I could always tell the managers who had the strongest leadership characteristics during those meetings because they were always the ones who were most willing to share. Others would worry that they might look weak. But the point is this is a learning environment."

Ilene Gordon, the CEO of Ingredion, which makes ingredients for the global food industry, started an annual tradition with young, high-potential managers at the company, bringing them in to make brief presentations to the board. The experience teaches them critical lessons about being concise.

"We have everybody give an 'elevator speech.' You have three minutes to tell the board and other people in the room where you came from, the challenges you're facing and how you're trying to create value for the company. Everybody can do that in fifteen minutes, but you have to be succinct," she said. "This is part of what we're looking for in people who have potential; it's all about communication. What are the challenges you have, and you have three minutes, because there are forty of you and we're going to be here all night, otherwise. And if you take somebody else's time, that's not respectful. It's all about being succinct and articulate."

If companies are going to invest more time and energy in training programs, there are some traps they should avoid, according to Karen May of Google.

"One thing that doesn't make sense is to require a lot of training," she said. "People learn best when they're motivated to learn. If people opt in versus being required to go, you're more likely to have better outcomes. You can influence people to come. If a group of people goes through some kind of program and they like it, then you ask them to nominate someone who might find the program beneficial. If the invitation comes from a colleague or a manager, you have that kind of peer-to-peer influence that says, 'I got something out of this. You might, too.' Then the people who come are motivated. They assume they're going to get something out of it. You just create a much different vibe than 'I was told I have to show up to this thing.'

"Another 'don't' would be thinking that because some training content is interesting, then everyone should therefore go through it. If something is interesting under particular conditions, it can lose its magic when applied to everyone. And don't use training to fix performance problems. If you've got a performance problem, there is a process to go through to figure out what's causing it. Maybe the person doesn't have the knowledge or skill or capability. Or is it motivation or something about relationships within the work environment? Or lack of clarity about expectations? Training is the right solution only if the person doesn't have the capability. But what I have seen in other places is sort of a knee-jerk reaction by managers to put someone in a training class if that person isn't performing well."

Lead and Teach

The tone of any organization starts at the top, of course, and creating a culture of learning—where school never ends—begins with leaders clearly signaling how much value they place on education. Steve Stoute of Translation LLC and Carol's Daughter said that one of his primary roles is teaching his employees.

"Teaching people who work for you is a very important skill set that requires patience," he said. "I've seen a lot of great leaders fail to execute because they couldn't get a team to rally behind them. You meet a lot of entrepreneurs who want to build great businesses and they have great ideas, but their leadership style doesn't allow them to have any patience to teach people."

Teaching is also at the core of the leadership philosophy of Dave Barger of JetBlue, an approach developed out of his own early experience as a young executive in the airline industry.

"Back in the eighties, I was at New York Air, and I was manager of stations and then director of stations," he said. "I can remember it like it was yesterday, because all of a sudden you have an awful lot of direct reports and pretty dynamic change. It was a shock to my system, because you didn't have years to really build the tools in your tool kit. At the same time, there was tremendous access to the leaders of the company, so you had great visibility. I think I've learned over time to try to expose as many potential leaders to situations as early as possible, because it really helps to build your experience for when you do move into that chair. As a young manager, to have access to the top of the organization was very memorable for me, because it helped set the framework at the highest levels about the behaviors that were needed and the goals that were being created.

"I think the best leaders are teachers and they're truly taking the time to explain a balance sheet or a fuel-hedging policy or other things. You're teaching. You're not just doing and communicating what you're doing—you're teaching people why you're doing it. And I really believe in giving people the opportunity to have access. There's got to be other people within JetBlue who can run this company who are not just my direct reports. They're in the organization and they've got great careers ahead of them. I also believe in leading from the back of the room, and watching people in the company who are stepping up to teach others."

Linda Heasley, the former CEO of The Limited, said that helping her employees learn new skills, which in turn helps retain talented staff, is at the core of her approach to leadership.

"I believe that my associates can work anywhere they want, and my job is to re-recruit them every day and give them a reason to choose to work for us and for me as opposed to anybody else," she said. "So it's about making it fun. It's about making it exciting. It's about keeping them marketable. I encourage people: 'Go out and find out what the market bears. You should do that and then come back and help me figure out what you need in your development that you're not getting, because we owe you that.'

"I've been told by my associates that's a countercultural approach to leadership: 'You're telling me to go look for another job?' But my point is that I should be able to re-recruit them. I should be able to get them convinced that this is the best opportunity for them. That's my philosophy."

12.

THE ART OF SMARTER MEETINGS

When two men in business always agree, one of them is unnecessary.

—WILLIAM WRIGLEY JR.
(1861–1932)

Many meetings seem designed to foster a culture that is precisely the opposite of quick and nimble. We've all been stuck in such meetings: there's no clear agenda, the discussion wanders, and people are constantly checking their smartphones and tablets.

The tug of devices is so strong that Seth Besmertnik of Conductor has taken a step to combat the problem of distracted meetings. "We have signs in every conference room in the office that say, 'We respect our colleagues by not reading e-mail during meetings.' This is one of the few things that drive me absolutely insane. Let's not meet if no one's going to be paying attention."

Why is a fundamental building block of work often handled so badly? Let's set aside that eternal mystery and focus on solutions. Many CEOs have smart strategies for keeping people engaged in meetings, focused on the task at hand, and clear about the agenda. They are also good at diagnosing—and avoiding—the common traps of meeting dynamics.

Keeping a Focus

Annette Catino, the CEO of the health insurance company Qual-Care, has a strict rule about meetings at her company.

"If I don't have an agenda in front of me, I walk out," she said. "Give me an agenda or else I'm not going to sit there, because if I don't know why we're in the meeting, and you don't know why we're there, then there's no reason for a meeting. It's very important to me to focus people and to keep them focused, and not just get in the room and talk about who won the Knicks game last night. That's not what it's about."

Agendas are important, but some CEOs take an extra step, clarifying the terms of engagement for a particular meeting so that people understand what kind of decision will be reached. Carl Bass, the CEO of the design software company Autodesk, explains his approach:

"One of the things we do is try to be very clear about decisions, because there's this built-in tension between hearing people's opinions and people thinking everything's a democracy. In some meetings, I will say up front that this is my decision and my decision alone, but I want to hear your opinions. We're very clear at the beginning of every meeting whether it's one person's decision, or whether it's more of a discussion to reach consensus. I think it's a really valuable thing to understand, because otherwise people can feel frustrated that they gave their opinions but they don't understand the broader context for the final decision.

"I think one of the main roles for a leader is to get as many opinions as possible on the table. But the flip side of that is you have to be clear when you're asking people for information and opinion but not turning over the ability to make the decision to them."

Sheila Lirio Marcelo, the CEO of Care.com, which helps

families find caregivers, developed a shorthand to signal the point of meetings at her company.

"We make three types of decisions at Care.com. We do Type 1, Type 2, Type 3 decisions. Type 1 decisions are the decision maker's sole decision—dictatorial. Type 2: people can provide input, and then the person can still make the decision. Type 3, it's consensus. It's a great way to efficiently solve a problem," she said.

Shorthand can also be used to avoid a common problem that many CEOs describe: they might wonder aloud about something, intending only to share a notion they're puzzling over, and their remark is taken as an order to be implemented immediately. Dawn Lepore, the former CEO of Drugstore.com, has a clever way to make sure people don't confuse the two.

"We have a little joke where I'll tell people, a lightbulb or a gun," she said. "A lightbulb means, 'This is just an idea I had, so think about it, see if you think it's a good one. Either follow up or don't, but it's just an idea.' A gun is 'I want you to do this.' People don't always know if you mean something as just an idea, or you want them to go do it."

It's not only important to make the purpose of a meeting clear at the outset. Bill Flemming of Skanska USA Building will set aside time at the end of a meeting to review the discussion.

"I've noticed in management meetings that you have to drive them to a decision and remind everybody about the decision," he said. "Sometimes people will try to change a decision after it's been made. You can't leave a meeting with the possibility that that might happen. So you have to have a little more formalized approach, and be clear about what was decided. I like to conclude meetings with two things. One is 'What do you think of this meeting? Was it time well spent?' And two is 'What's your commitment to this team? What are you going to commit to doing when you leave this room?' In other words, that's the personal commitment to your group when you leave. It's time well spent. I'm not saying we're a hundred percent successful every time, but I think that's a good way to conclude a meeting. I find that with business meetings, if you don't pay close attention, they can just wander."

Finding a Positive

Setting the right tone at the outset can make for a more productive meeting. Ken Rees of Think Finance asks his staff to start each meeting with one positive piece of news.

"We do a daily huddle, and we always start out with good news," he said. "If you're trying to change a lot of things, and you've got challenges, sometimes people can get pretty beaten down. So we do try to start things off with one piece of good news from each person. And there's always one person who says, 'I can't think of any good news.' And we say, 'No, you've got to come up with some good news.'

"In an environment that's changing as quickly as ours, there are so many things to be worried about. And it's easy to get focused on the fact that you're not where you want to be, as opposed to remembering all that you've accomplished. So taking a moment to talk about good news just keeps things more upbeat than they otherwise would be. For a lot of CEOs, I think, we're so focused on where we need to be that we don't necessarily have the fun we need to have along the way."

Amy Gutmann of the University of Pennsylvania encourages brainstorming and pushes people to bring up even far-fetched notions.

"I love challenges and I'm enthusiastic about taking them on with a team," she said. "And my team knows that I like good ideas even when I disagree with them, and that I'm hard-driving but also reward everybody on the team who combines passion, smarts, and hard work. I've also written a lot about the importance of deliberation, and we practice it. We bring everybody to the table and I like to say, in any given week, 'This is my wild idea for the week.' We don't execute more than half of them, but for the ideas that we put into practice, everybody was at the table and I think they get the same kind of excitement and satisfaction that I get out of it.

"What I've learned over time is that while you're driving all of your priorities forward, it's really important to get feedback and to be open to the wild and crazy ideas, even if you're not going to pursue but a fraction of them. And that probably makes a lot of sense for a university, because we are all about ideas. If we're not open to them, if I'm not open to them, who is going to be? I encourage other people to do it and we're not shy about shooting ideas down. If it's intended to be wild and crazy, most of them are going to be shot down. But the ones that survive, we all rally behind. And, yes, I definitely expect other people to outdo me. I think people on my team recognize that I am very straightforward.

"When I believe that something is absolutely right and we have to do it, I don't spend a lot of time deliberating about it. I just say, 'We've got a problem here and we've got to solve it and tell me how to do it.' When I have a wild and crazy idea, I want them to know that I have no idea whether we can run with it, so tell me what you think and be as straightforward as I am about it."

When a leader proposes an idea in a meeting, staffers may simply nod and tell her how smart she is. Kyle Zimmer, the CEO of First Book, a nonprofit that provides books to needy children, gets around this problem with a clever technique.

"In a recent meeting, I had two things I wanted to float," she said. "I said, 'Now somebody has to stand up and tell me three reasons why this is sheer genius.' Everybody's laughing, and then I said, 'Now I want people to stand up and tell me the three reasons why this is the dumbest thing they've ever heard.' It's playful, but it puts people at ease and allows them to say things that maybe they wouldn't otherwise. It works. It pushes critical thinking, and it does it in a way that's not oppressive.

"I started using it with clients and corporate sponsors. If something has happened that's not great, it's very hard for a corporate person to say that, because they feel as though they're undercutting the Home for Widows and Orphans. They're sitting on the corporate side with all the leverage, and they know that. But if there's anything that has gone wrong—and after twenty years of having these relationships, of course stuff doesn't go perfectly every time—you have to fix it. And the only way to get them to tell you is to say, 'You've got to tell me the three things you love about working with First Book, and you've got to tell me the three things, if you had a magic wand, you would fix, or you would change.' It not only puts them at ease; it also gives First Book feedback that allows us to refine everything we do. That's a huge gift. But it's just a gift that they feel awkward giving sometimes."

Disagreements and moments of tension are to be expected in meetings, especially in cultures where candor is valued. Geoffrey Canada of Harlem Children's Zone has a technique for defusing such situations.

"There's not a day that goes by that I don't draw on my undergraduate background in psychology," he said. "I just don't understand how people manage well if they don't know anything about any of these underlying dynamics in these group settings.

"You've got a certain competitiveness of all of the people sitting around a table who all want to show how brilliant they are, and how helpful they are. You've got people who are absolutely going to be defensive, because maybe they're responsible for the thing that's being pulled apart right in front of their eyes. I always love this technique of just saying to that person, 'If I were you right now, I would just be feeling like everybody's attacking me, and wondering why I volunteered to do this.' Then people start laughing, because they know that's exactly how the person feels. You pull down the level of anxiety, and you create some transparency so that people realize the point is actually how you solve problems, and not the other dynamics of the meeting."

Drawing People Out

Anyone who's ever run a meeting knows how difficult it can be to strike the right balance between being in control of the meeting and encouraging everyone around the table to speak up. It is an art that requires constant practice, even for experienced leaders, as Amy Schulman, an executive vice president and general counsel at Pfizer, explains:

"One of the biggest lessons I'm learning now is having a better feel for when to step out of a situation and when to step in," she said. "I do think that is actually one of the hardest things to balance correctly. People want to hear from you. They want your opinion. And if you don't ever speak up and weigh in, then I think the people you lead will feel frustrated, wondering why you're hanging back and not saying what you think. But if you're constantly giving direction and speaking, then you're really not encouraging conversation. And no matter how democratic you'd like to think you are as a boss, you learn that your voice is louder than others'. I respond best to people who challenge me, and I like being challenged, and I tend to reward people who are appropriately challenging. I think learning to refrain from speaking—without making people feel that you're trying to frustrate them by being opaque—has been an inflection point for me.

"I learned that just by watching the room, and being puzzled if

I thought there should be conversation, and wondering why there wasn't more conversation. I also saw how quickly people tended to agree with me, so I thought, 'It can't be that I'm right all the time.' And so I learned to really try to deliberately reward people in a conversation for challenging me. I don't mean being insubordinate. I mean really following up on other people's ideas. One of the marks of a good speaker is actually being a great listener. So I remind myself that no matter how quick I think I am, that I have to show that I'm listening, and show people how I've gotten to the endpoint, or else I run the risk of squelching conversation. So I will deliberately slow myself down so that the room catches up to where I am. I know how I feel when I get cut off, and so shame on me if I do that to other people."

How do you get people to contribute, to speak up during meetings, to make sure that it's a vigorous discussion, and avoid those awkward moments of silence when people are asked to share their opinions? Several CEOs said an explicit part of their culture is that people are expected to offer opinions.

"I want a really authentic, good conversation in meetings," said Jenny Ming of Charlotte Russe. "We had interns this summer, and in the beginning they were afraid to speak. But in the meantime, they are our customer—I mean, literally—because they're twenty-one, twenty-two, and that's who we sell to. So in the beginning, they wouldn't say how they feel. And I said: 'Tell me what you love, what you don't love. You have to have an opinion because actually I pay for your opinion. I pay you to have a point of view, good or bad.' And toward the end of the summer, they couldn't wait to tell me what they thought. Most people can't wait to tell you what they think anyway, but they didn't know that they could.

"I've always done that, even when I was in Old Navy and Gap, because I think that kind of environment is very important. We get probably the best advice or the best decisions when it's a little bit more democratic. I don't always have to agree with them, but I certainly should listen to them because they are more of the target customer than I am. I think everybody has to have a point of view. If they don't have a point of view, they shouldn't be sitting around the table, because it's about clothes. Do we believe in this? Do we like this? Should we go after this? How can you not have a point of view?"

Romil Bahl, the CEO of PRGX, a data mining and auditing firm, is also explicit with his staff about expecting them to speak up during meetings.

"When I joined PRGX, I entered a pretty siloed and not very happy place, and so I had to break those things down," he said. "And so I stressed collaboration and teamwork. I stressed that the power of information is in sharing it and not hoarding it. In my leadership meetings, I ask my team to do what I try to do, which is lead from the front. And I say: 'It starts with you challenging me in this room, challenging each other in this room. When we leave the room, we've got the one way we do things, but in this room, let's get it out and please go encourage your people to do the same.' It's the best of the company that has to be brought to bear for everything we do. And the best idea has to win. It's not a sign of weakness to ask for help or to bring other people in. It's a sign of strength.

"People are used to me saying things like, 'Okay, for twenty-five minutes I haven't heard anything from you. What's going on over there? Are you with us? Are you looking at your BlackBerry? What's going on?' So it's a little bit provocative, as well, and you have to be careful about how many times you do that. But I'll do it because I have this fundamental belief that these are successful, great people around you who are here for a reason and they have something smart to say. At least on the most important things that we're going to do, until I feel like we've really hit upon that best idea, I'll keep pushing people."

Alan Trefler, the CEO of the business technology company Pegasystems, said that one of the core values of his firm is designed to get everyone to share his or her opinion.

"When people ask what the company is like, I say the culture we try to encourage is a 'thought leadership' culture," he said. "You hear people throw around that phrase a lot, but to us, thought leadership means some very specific things. We focus on each of the words. So, you have a thought when you have an opinion about something. You actually need to have an opinion that is hopefully a unique or complementary opinion to the opinion of others. As William Wrigley Jr. said, 'When two men in business always agree, one of them is unnecessary.'

"I think having an opinion is important, but it's not enough to have an opinion—it has to be an informed opinion. So content really matters, and you need to understand the context of what you're trying to have an opinion about. And then the second part of the phrase 'thought leadership' involves the concept of, 'What does it mean to be a leader?' And ultimately, you're only a leader if somebody's willing to follow you. And the characteristic about leadership that we focus on in that context is persuasiveness. As a company, we don't rely on formal authority. I often tell people that if somebody cites me as the reason for doing something, they should throw them out of the office. If I want somebody to do something, I will personally find them and tell them. And, short of that, people should do things because they've become persuaded or because they end up sharing the opinion of somebody else that it is the right thing to do. The people who know things should get to make decisions."

Drawing people out not only enriches the discussion with more opinions, but it also avoids a common and insidious dynamic, where some people stick their necks out and share their opinions while others simply sit back and critique. Julie Greenwald of the Atlantic Records Group has a particularly low tolerance for people who try to hang back.

"In meetings, I constantly talk about how we have to be vulnerable, and that it's not fair for some people in meetings to just sit or stand along the wall and not participate. If you're not going to participate, then that means you're just sponging off the rest of us," she said.

"I'll throw out ideas. Some of them will be horrible, and I'll let people to the left of me or the right of me gong me and tell me that was the wackiest idea on the planet, and we'll get through it. They won't get fired. And then I'll say to others, 'Okay, you, what's your idea?' It's important for everyone to understand we're a company where risk taking is necessary.

"I know it's not easy sometimes. I hate public speaking. I hate it, I hate it, I hate it. The only way I conquered it was being put on the spot all the time. In order to lead, you have to be a public speaker, and in order to be a department head, you need to be able to really drive the meeting and make sure everyone understands that the reason you're the department head is because you're a driver. You're not just someone who's taking a backseat."

Kathy Savitt of Lockerz captured this dynamic in a memorable phrase.

"Like a lot of people, I've worked in companies before with what I call a 'stump a chump' culture," she said. "It's very toxic. 'Stump a chump' is when a CEO or a leader will throw out a question and somebody will offer an answer. But nobody else has the courage to answer, and yet, everyone then critiques the answer that the one person offered. Or you do something creatively and then everyone else in the company calls out what they would change, yet they haven't put either the professional or the personal passion behind actually creating something of value. As a leader, you're not only an amalgamation of the best practices you've seen, but also an amalgamation of the things that you've loathed in other organizations."

Rethinking the Meeting

Some leaders avoid the common traps of meetings by taking a different approach. Tim Bucher of TastingRoom.com finds that a weekly meal for his executive staff is a more effective way to get people to share ideas.

"Staff meetings are important," he said. "But they're usually tactical: 'What happened last week that we don't want to have happen this week, and what's got to happen this week?' But at these kinds of staff meetings, you see only two months out, if that. I've also seen companies where, in addition to staff meetings, they have an off-site or a leadership summit every quarter or maybe once a year. Isn't it amazing how great those off-sites are, and all these ideas come up? Then you go back to work, and there's zero follow-up. It's such a huge moment for the team, and they get all excited, then they get back to the real world, and it's like nothing ever happened.

"So I have a COO who runs the weekly tactical meeting. But I take my team of eight people to dinner every Monday night. I do it because we cover the tactical stuff in the staff meetings, and we cover only strategic stuff at dinner. I think that when you are in that kind of setting, you can talk more openly about things like new products and opportunities, and structural changes in the organization. We get into some pretty interesting conversations. It's not that I think traditional staff meetings are pointless. It's just that I think that most times you can never get anything strategic done. Even if you have a four-hour staff meeting, you're just not going to go into the strategic stuff. After all, it's during the day, people are texting and e-mailing, and nowadays everybody comes into a staff meeting with their computer, and they open it up. There are no computers at dinner. There are no interruptions. You get a lot more honesty. The other thing about it is that it's like having an off-site every week."

Amy Astley, the editor in chief of *Teen Vogue*, said she prefers small groups to larger meetings.

"I try to create a culture that's very receptive to ideas from everyone on the staff," she said. "But I've learned something about meetings, by trial and error. I used to do big staff meetings, with everybody sitting in the conference room. After I did that for six months or so, I realized that it quickly becomes like a high school cafeteria. You have your alpha girls. Two of them are best friends. They talk. They shut everybody else down. The other people don't say a peep—you'd think they've gone mute—and no new ideas are coming out. So I'm very strategic about how I meet with people. I have an open-door policy. I basically spend my day meeting with my staff, one or two people at a time. People come in and out all day long. We talk about ideas, and nobody feels attacked.

"I find a lot of meetings tend to be recaps. And it's not always clear who's going to do the work if there are too many people in the room. If I meet with one person, they know that they are accountable. We discussed it. I told them what I want them to do. They tell me how they think it should be done. Then they go do it. It's very clear if they didn't do it. With twenty people in a room, it's too easy to just pass it off. Smaller meetings are also better because people are more open. They'll put things out there, and more ideas come out in one-on-one meetings because it's very personal."

Get to the Point

Nothing drains the energy out of a meeting faster than somebody who launches into a long-winded explanation. The people who have to sit through these speeches may start silently reciting the old saying about wanting to know what time it is, not how a watch works.

The trick, when asking people for their opinion, is to signal to the room that people are expected to be concise by framing discussions with phrases like, "What's the ten-word pitch on that?" Catherine Winder, the former president of the animation studio Rainmaker Entertainment, sets a clear expectation that people need to be brief when they are proposing an idea.

"We want everyone to think of themselves as storytellers, so we have an open-door policy for anyone in the company to pitch ideas," she said. "Our main initiative is encouraging staff to develop ideas for stories using the characters we created as our company mascots, and to describe their ideas to us in fifteen to thirty seconds. If we like the core idea, we'll work with them to develop the story so that the short can be produced. If you can be concise and come up with your idea in a really clear way, it means you're onto something. In my experience, if you give someone a minute, you're going to get five minutes."

Joel Babbit, the CEO of the Mother Nature Network website, always keeps his meetings short and to the point.

"I probably run the shortest meetings. Sometimes it's just a yes-or-no answer and then I walk out," he said. "I have sat through some excruciating meetings, and mine are probably too short. But at the same time, I believe that the majority of meetings could easily be cut to a third of what they are and accomplish much more. People just lose track after some point, and it doesn't matter what's being said. If you go into a meeting where somebody's talking for more than fifteen minutes, ninety percent of the people are looking at their BlackBerrys and not paying attention.

"I've also found that the higher up you go in a corporation when you have a meeting, the shorter the meetings are. So if you meet with somebody who's an assistant vice president, the meeting will be two hours. You meet with a vice president, that's an hour. You meet with the chief financial officer, it's a half hour. You meet with the CEO, it's ten minutes. That's how it works."

13.

KNOCKING DOWN SILOS

I firmly believe silos ultimately destroy a company.

—Chauncey C. Mayfield,
CEO of MayfieldGentry
Realty Advisors

As companies grow, tribal behavior inevitably kicks in, and people align themselves in smaller groups. They start talking about "us," meaning their part of the organization, and "them," meaning other groups within the company. It's a phenomenon that Mark Fuller of WET Design describes as "these fiefdoms that tend to get sclerotically reinforced over time in companies when people say: 'Oh, the fifth floor is engineering. You don't go up there without a hall pass.'"

As discussed earlier, shared goals are an effective way to encourage employees to see more people as part of their "in group." But many CEOs take additional steps to knock down the silos that naturally develop in an organization. Fuller took an unconventional approach. He brought in an improv drama teacher to encourage people across his company to listen more carefully to one another. He explained his rationale:

"We have three classrooms and a full-time curriculum director who teaches all the time and also brings in outside instructors. One of the really fun classes we do is improv. If properly taught, improv is really about listening to the other person, because there's no script. It's about responding. I was noticing that we didn't have a lot of good communication among our people.

"If you think about it, if you have an argument with your wife or husband, most of the time people are just waiting for the other person to finish so they can say what they're waiting to say. So usually they're these serial machine-gun monologues, and very little listening. That doesn't work in improv. If we're on the stage, I don't know what goofball thing you're going to say, so I can't be planning anything. I have to really be listening to you so I can make an intelligent—humorous or not—response.

"So I got this crazy idea of bringing in someone to teach an improv class. At first, everybody had an excuse, because it's kind of scary to stand up in front of people and do this. But now we've got a waiting list because word has spread that it's really cool. You're in an emotionally naked environment. It's like we're all the same. We all can look stupid. And it's an amazing bonding thing, plus it's building all these communication skills. You're sort of in this gray space of uncertainty. Most of us don't like to be uncertain—you know, most of us like to be thinking what we're going to say next. You get your mind into a space where you say, 'I'm really enjoying that I don't know what he's going to ask me next, and I'm going to be open and listening and come back.' We've got graphic designers, illustrators, optical engineers, Ph.D. chemists, special effects people, landscape designers, textile designers. You get all these different disciplines that typically you would never find under one roof—even making a movie—and so you have to constantly be finding these ways to have people connect. So we do things like improv, and I think they really have developed our culture."

Of course, improv classes may not be the right fit for many organizations. But CEOs know they have to do whatever it takes to knock down silos.

Create Incentives

Compensation and reward systems are set up to encourage certain behaviors; leaders can use their organizational charts to send powerful messages about discouraging siloed behavior.

Chauncey Mayfield of MayfieldGentry Realty Advisors has implemented a performance review system that reflects employees' work across his organization, not just what they do for their immediate supervisor.

"What we're finding, and I've found this in other companies, is that people start to build silos and the only thing they really care about is what's going on in that silo," he said. "So this silo is not working with that silo, and now you have almost warring factions within companies. They have no desire to work with each other, because there's no downside for not doing it, and you're not incentivized to do it. So, as we've gotten bigger, I've been working to try to make sure those silos aren't built up, because I firmly believe silos ultimately destroy a company. That is probably one of the biggest challenges I face—how to make sure these silos aren't built up.

"We've kind of redesigned our approach a little bit and it's called a customer service approach. It's one thing to keep your boss happy, but it's another thing to keep the colleagues you're also working for—your customers inside the company—happy. People tend to manage up, but you have to manage across as well. And so what it's fostering is people working together. It helps foster a team approach, knowing that that guy over there also has a say in my bonus, not just one person."

Many firms measure performance at the individual and office level. Martha Samuelson of the Analysis Group took a different path.

"We have one P&L [profit-and-loss statement] for the whole firm, even though we have ten offices," she explained. "We don't have practice-area P&Ls, and we don't have office P&Ls. We don't even have sales credits for the partners. We have a trust-based system for setting partner compensation, and it's based on a belief that we're in a long game together.

"I was at another organization before I came over to the Analysis Group. They had office P&Ls, and there would be times when one office was working flat out and people in another office were twiddling their thumbs. It made a very deep impression on me. I thought it was awful. If you remove as many disincentives to cooperating as you can, even ones that seem like they would be useful, you find everybody cooperates much more. That said, it's not for everybody. There are different ways of running companies. Different groups of people are going to care about certain things more, and in my organization people care about getting along and teamwork."

Samuelson said she understood there was some risk of people feeling like they were not being fully rewarded for their contributions. But such a system, she said, also attracted certain kinds of employees.

"I think people here are willing to trade off the fact that there may be measurement error," she said. "So it may be that this year, with the best intentions in the world, somebody gets underpaid a little bit compared to a system in which you were measuring every single thing. But we're going to be here for a long time, and on the whole people just like it more. We're not a star system and we're not going to take the company public. The people who have stayed and thrived have been people for whom this collaboration issue is so important that they're willing to leave some money on the table over it. It is for me, certainly. Work is really important to me, and I feel like I've got one trip through the planet and I would rather do it in a certain way. I feel like, on the margin, more money is great. But if you're trading off a work environment that is really working well for you, it's not worth it."

Kathleen Flanagan of Abt Associates reworked the organization chart when she took over her consulting firm.

"As the company has grown, more business units were created, and so we had more silos in the organization," she said. "My objective was to take down the silos. So I reorganized the company. It used to be organized around lines of business—international, U.S.-based, data collection—and there used to be senior vice presidents who led each of those big businesses. I took those senior VP positions away and hired one executive vice president for global business who shared my vision for what I call One Global Abt.

"At the heart of that is taking down the walls so people can collaborate more freely, so that we can leverage all of Abt. We now ask people to pick their heads up out of their project work or their division focus and look across the whole company. So I now ask my managers to wear two hats. Everybody's got their job in the big picture of the company, but they all have to wear an Abt hat. It's really easy, given the time pressures and the pace of our work, to put blinders on and be very project-focused. It's harder to take a step back and ask, 'How does this apply to the whole company?'"

Bring Everyone Together

Many companies hold regular all-hands meetings, bringing the entire company together in person or—if they have far-flung operations—through videoconferencing and conference calls. They are particularly effective for combating a problem that Geoff Vuleta of Fahrenheit 212 touched on earlier in the book. His point is worth repeating here: "One of the truisms about life is that if you're working in a void for any period of time, human nature says you'll view it negatively. You get scared; you begin to believe that what isn't there is probably bad. Never give people a void. Just don't, because instinctively they'll think something is awry."

Jeremy Allaire, the CEO of the online video platform Brightcove, recognized the need for holding town hall meetings during a period of particularly rapid growth at his firm.

"We grew to a size where I was feeling kind of out of touch with everybody, and that felt a little bit scary. We went from a hundred and eighty people to almost two hundred and eighty in less than a year, so that's a lot of change," he said. "I actually was inspired by one of the rituals the founders of Google established: Every Friday they would invite anyone in the company on a worldwide basis to join a town hall to just talk about anything that's happening in the company.

"So every Friday at ten a.m., we tell people, you've got an hour and we're going to talk about anything. Everything is on the table, and we'll bring some topics, too. It is completely open book, and people can ask any hard question. Not everyone shows up, because people have a lot going on. But it just creates this sense for people that they've got access to anything that's going on in the company strategically. It's a really helpful thing."

Laura Yecies of SugarSync saw the value in expanding the all-hands meeting, on one occasion, to include employees' families.

"One thing I learned in my first year at SugarSync—when the economy was bad, people were nervous, and we didn't have a lot of money—was that I had to sell to the families, not just to the employees. The employees had more information, of course, because I was talking to them frequently and they knew that we have this product we were excited about. They could have a sense for the market, but their spouses or parents or significant others didn't have as much information. So we started doing more to communicate with them. In fact, we had a SugarSync evening where we invited people to bring their spouses and their children. We had a babysitter and videos and pizza, and we did a demo for the spouses and significant others, whoever people wanted to bring, and I gave them our investor pitch. It turned out to be a really good event."

Susan Credle, the chief creative officer of the ad agency Leo Burnett USA, had an interesting first-day speech to remind new employees that the competition was outside the building, not among staff members.

"Take out your business card and look at it," she told her staff. "That business card will have more value if any one of you succeeds here, even if you're not remotely a part of that success. You are not competing with each other in here. If you think you win when your idea wins out over your neighbor's, that's a pretty small gain. In fact, I would suggest that you help your neighbor's ideas get better.

"I would suggest that if you look at something and you have a better idea, that you generously give that idea to someone and make them better. Because if we all do that, we all win. The minute you're the only good thing at this company, we're done. So can you do it? Can you be that generous? Generosity has a lot to do with confidence. If you're confident in who you are, you will be generous. If you're scared, if you're nervous, if you think you're a fraud, you won't be generous."

Daily breaks can also help knock down silos.

"The small things matter," said Doreen Lorenzo of Frog Design. "We have many rituals. One is something called coffee time: every day at four o'clock, in every one of our studios around the world, everybody stops and they have the opportunity to go into the kitchen and people just socialize. They might play a game of Ping-Pong, they might play a video game, and there are pool tables, foosball.

Different studios have different toys. That's a ritual and that's just accepted. That's what we do.

"These are intense people. This is a time for them to take a break, to talk to people they might not work with, and to listen to things. That's every day, Monday through Friday. We often joke that if we ever took coffee time away, we think everybody would quit. And we have ten o'clock Monday morning meetings at every studio, where they go over anniversaries, birthdays, projects, and share stuff that another studio has done. That's something that's really important to people. Then each studio has a little bit of its own culture to inspire creativity. They might go on field trips. There's a 'sketch jam' in one of the studios every Tuesday at noon, and people come in and they just practice their sketching skills. But that's all accepted. That's part of the culture."

Assigned Seating

Yes, the notion can remind people of elementary school. But forcing people to meet and work with others, rather than hunker down in their offices and cubicles by themselves, can help keep silos from forming.

Sheila Lirio Marcelo of Care.com has made shifting seats a tradition at her company.

"Everybody moves around every year. And people don't have a choice where they sit; we rotate them. Part of the reason was to embrace change, to remove turfiness so that you're not just chatting with your friends and sitting with your friends. You sit with somebody else from a different team so you get to know their job. What are they doing? What are they saying on the phone? How do they tick? And it's getting to know different people, so that we build a really big team. And we do that every year. And it's now actually become an exciting thing that people embrace."

F. Mark Gumz, the former CEO of the Olympus Corporation of the Americas, set a rule during his tenure that people weren't allowed to eat their lunch at their desks, but, rather, had to go to the cafeteria. "I want you to take a break and talk to other people in the company," he said.

He also created ways to introduce people to colleagues from other areas of the company, to create a sense of community.

"You have to have an open management system," he said. "And so our quarterly town hall meetings are critical. Good times or bad, you tell the story—people can deal with the information. You have coffees. You have lunches. You sit in the cafeteria, and move table to table. You encourage your team members to do the same. You have management meetings, and you have tables where you tell people where they're going to sit; they don't sit with the people they work with, because you want them to meet new people.

"We also do volunteer activities and send our people out into the community. We put teams together, and you don't know who you're going to be working with. And it's incredible what happens when people start to talk to other people in the business. We're starting to see people moving around in the company—that's how you build careers for people. I think that's how a company becomes stronger and stronger, and that's how passion is developed. It's not a bad thing to give people an experience to take them out of their comfort zone. And if you do it often enough, people feel comfortable with trying new ideas."

Meeting spaces themselves can have a powerful effect on the discussion. If a small group is spread out around a big table, people can feel distant from one another—literally and metaphorically. Julie Greenwald, of the Atlantic Records Group, will often bring people together in tight spaces.

"I'm not afraid to call a meeting and shove seventeen people into a tiny office. We look like a clown car. But you know what?

It's okay because that's when you feel like, 'All right, we're this tight-knit unit.' So I started to assemble a whole other set of meetings in my office, and they're like SWAT team meetings, and somehow they've gotten bigger and bigger," she said.

"Sometimes I look around and I think, 'Am I an idiot that I've got fifty people crammed into my office?' But it's so tight and we're so on top of each other that you feel that 'It's us against the world, man.' We're not in the conference room. And everybody's on top of each other, so there's no seating chart. All of a sudden it sets a whole new playing field where there's no hierarchy. Everybody's now on equal footing. It's like a free-for-all."

Draw on Design

Encouraging people to interact more is not just a matter of giving people assigned seats or moving them around to new desks. The office layout can also have a big impact on the way people work together. Michael Lebowitz of Big Spaceship made some changes when he noticed something was not quite right.

"When we moved into the space we're in now," he recalled, "we went through the process with the architects of talking through what we wanted the space to look like, and they did these drawings of these really awesome, sort of monolithic giant white tables. I think there were seven of them across, and they can seat about ten people around them. And there was a point after a few years when I could feel a little territorialism creeping in: 'Oh, that wasn't us. That was the technology group.' And I said I need to nip this in the bud. We were sitting according to discipline. As soon as they're sitting together, it's a department.

"So I worked with my core team to figure out a different layout, with interdisciplinary teams seated together. And we changed the layout so that rather than facing each other, they face outward. They can focus on their job, but as soon as something needs to be discussed, all they have to do is swivel their chairs in and they've got a team meeting. And it's wildly efficient. We have teams that have no conference-room meetings at all and are very, very organized."

It's easy for people in different offices to feel disconnected from other groups. The sheer distance between the offices can create silos. Some CEOs of tech firms have started taking advantage of the falling prices of massive television screens to create "video walls." Phil Libin of Evernote explains how they work:

"So our headquarters are in Redwood City [California], and we set up a studio in Austin, Texas. We very specifically wanted to avoid the feeling that if you're not working at headquarters, you're in a second-place office. So in two high-traffic areas each office, we have this giant seventy-inch TV with a high-end video camera on it, and it's meant to be, basically, a window. So they're always connected in real time. The idea is that if you can see somebody, they can see you, so you don't have that feeling of surveillance. It's really about connection. You can see who's on the other side and talk to each other. But it's not for meetings—this is really just to connect the spaces. And you can just chat with each other, and maybe that will trigger ideas."

Dennis Crowley of Foursquare, who set up a similar connection, said it paid immediate dividends.

"I started to feel a bit disconnected from our San Francisco office," he said, "so we got two big screens with cameras there and here in New York. They're on all day long, so you just walk by and say, 'Hey, Pete, what's up? Can you get Ben?' It works so well."

Brent Saunders of Bausch & Lomb took a more dramatic step shortly after he arrived at his job: he called in the architects to move the executive offices out of a traditional headquarters building.

"I think changing a culture requires multiple actions, and actions speak louder than words. So you can talk about culture all you want, but it takes a while to seep in because it's really about what you do," he said. "We owned a skyscraper in Rochester, but we had close to a million square feet a few miles away, where we had a manufacturing plant, R&D, customer service, sales and marketing, all under one roof. Yet all the executives were sitting in this fancy tower, and everybody who really did the work was sitting in this other facility.

"So I walked in after four weeks on the road and saw my huge office with this skyline view on the top floor of this building, and I said, 'This isn't going to work. If we're really going to create one company, one culture, one team mentality, then we should all sit together.' And so we moved all the executives out of there to the building with everybody else. It was a great symbol that there's not going to be this bifurcation of culture between the people sitting in the big tower and the people actually doing the work. We're all in this together, so let's all sit together.'"

Put on a Show

In the grind of day-to-day work, it can be easy to start taking the talent and skills of colleagues for granted. Some CEOs create occasions for employees to be able to sit back and admire the work of the organization, and the people who work there.

Susan Credle of Leo Burnett USA established a tradition at the agency of doing an internal awards show every year. The entire staff gets to see all the work their colleagues have done.

"We set it up like Cannes. We invite every employee to come and vote on the work," she said. "This year, they had three days to vote, and we had over a thousand employees vote. It's a great experience, because half of our employees didn't know about everything we do. But then, all of a sudden, once they realized it, they started walking with a little more swagger. I would say to them, 'You're a part of it all. Because whether you say something inspiring on the elevator or you're just nice or you put some positive energy into this office, that's all helping us make that work.'

"We started it the first year I arrived, because I was supposed to put a reel together of the work and I wasn't thrilled with it. I thought, 'If I stand up there and go, here's what you did this year, then I'm actually accepting it and saying this is fine.' So I decided that if we all voted on the work, then it's everybody saying 'This is what we did this year, and this is what we think is the best.' So it took it off of me, and became a companywide process of commenting on our work. We realized the benefit was phenomenal. The other thing is that we broadened who we recognized for the work. Traditionally, it's usually the copywriter and the art director, and maybe the creative director, who are recognized. But we recognized everybody on the team, including who does the budgets, who does the financing, and they all get listed."

Although a circus may seem like a far-fetched option for more traditional companies, there are insights to be gained from Kenneth Feld, the CEO of Feld Entertainment, which produces shows like the Ringling Brothers and Barnum & Bailey Circus and Disney on Ice.

"Every year when we do a new show," he said, "we bring down to Florida a hundred and thirty performers who never, for the most part, knew each other beforehand. About ten days into our rehearsal period, we have Act Night, where everybody performs their act uncut, untouched by us, for every other performer.

"That is the true test of respect. No matter what anyone thought of a person, if they're doing an act that is so unbelievable and death-defying, the respect level goes way up. You have earned the respect strictly by what you have done—it's very pure. It is an absolute lesson in earning respect. Respect does not come from a title. It comes from what you do, and how you do it, and how you work with people, and I think that's a difficult thing for people to understand."

14.

SPARKING INNOVATION

As a manager, you have to be able to entertain chaos. Otherwise, you'll never get close to the creative process.

—CHARLOTTE BEERS, former CEO,
Ogilvy & Mather Worldwide

Ronald M. Shaich, the chairman and co-CEO of Panera Bread, has developed a simple framework for thinking about how companies operate. He says that organizations really have two muscles: a "delivery muscle," for getting work done, and a "discovery muscle," for trying to innovate. But most organizations take a lopsided approach, as he explained:

"We have not fallen prey to what happens in so many large companies: they let their 'delivery muscle' completely outweigh the 'discovery muscle.' The delivery muscle, of course, feels rational, people feel much safer with it, and you can analyze it. It's driven by market research that tells me what I can count on, and it's very good for incremental change. But when you are talking about companies that find new patterns, that have discovery, it's about leaps of faith. It's about trusting yourself. It's about innovation. It's about believing you can figure out where the world is going and that you will get there, and then saying we are going to go left and not right. The decisions are not necessarily rational or easy to defend. Even so, you need to take that approach."

All company leaders want their firms to be more innovative. They see innovation as a key to revenue growth, particularly as the economy continues to struggle. They recognize that their competitors are pushing hard to innovate, and that they can be easily left behind if they don't develop new products and services that customers want and need. Innovation can also cut costs, if more efficient ways to operate are found. Companies can't afford to let their discovery muscle weaken.

But here's the good news. People throughout organizations often have plenty of ideas. They're on the front lines, seeing how things work, how they don't work, and how they can be made to work better. They hear customers describing problems they would love to see solved, and they notice points of friction and inefficiency inside the company. The role of a corporate leader is to create a culture that channels those ideas, that energy, into solutions that matter. Because that's the difference between invention and innovation—people can invent all sorts of things, but doing so counts as innovation only if the result is a product or service that a customer is willing to pay for.

Brent Saunders of Bausch & Lomb found that employees at the struggling company had plenty of inventive ideas when he was brought in to turn the business around. The challenge was to focus this energy in a way that led to more innovation.

"One of the things I learned early on was that we hadn't brought a lot of meaningful innovation to market in four decades," he said. "We had some innovation but not the kind of innovation that really would make a difference in people's lives or for customers. I looked at our R&D organization and I saw that the talent level in that group was really high. It dawned on me that these are really talented people who are really focused on getting patents and publishing papers and creating process around product flow, but not really getting product out the door. And so when I started to talk to them, I learned that what they really wanted was what we all wanted, which was to create something in their labs that helps somebody and that would also help the company.

"I think the disconnect was from a lack of focus on what success was. Success wasn't around how many patents you had or how many papers you published. Success needed to be defined as creating products that mattered and made a difference for patients and customers. And so one of the ways we did it was by a semantics shift from 'R&D' to 'D&R,' to really show people that while we still invest in research, let's prioritize the development side.

"A lot of scientists and engineers didn't like the shift at first. But they grew to love it because you can come up with all these wonderful ideas and patents and papers and scientific findings, but if you don't actually get them out of the lab and turn it into a product, you're not helping anybody and you're not helping the company."

To put an even sharper point on the difference between invention and innovation, Steve Case of Revolution likes to quote an expression that he attributes to Thomas Edison: "Vision without execution is hallucination."

"I do believe in vision," Case said. "I do believe in big ideas. I do believe in tackling problems that are complex and fighting battles that are worth fighting and trying to, in my case, create companies or back companies. The vision thing is really important but the execution thing is really important, too. Having a good idea is not enough. You've got to figure out some way to balance that and complement that with great execution, which ultimately is people and priorities and things like that. You have to strike the right balance. If you have those together, I think anything is possible. If you don't have both of them working together in a complementary, cohesive way, you're not going to be successful."

Feed the Monster a Cookie

Terry Tietzen, the CEO of Edatanetworks, which develops customer loyalty software, uses a simple system for breaking down innovation into more manageable steps, to avoid some of the political dynamics in groups that can distract and derail progress.

"You have to be careful about gab sessions where you ask people to give their opinions," he said. "The downside about opinions is everybody has one. You can open up a Pandora's box. You destroy innovation sometimes by asking for too many opinions, because people can feel dejected and put their head down if their idea doesn't get accepted right away.

"So you avoid that by saying, 'Listen, I have this vision right now. I'd like to share with you how I think it will work. Then can all of you do me a favor? Go work on it and tell me what you think at the end of the day.' I get their buy-in by working on it rather than asking them their opinion up front. Otherwise, it's a tug-of-war all the time. Then they start giving me daily updates on what they got done today, and what they're going to do tomorrow. And in the planning for tomorrow, they give you the great innovation before their opinion. Because opinions can cause tension—people tend to compete over whose opinion is better—and you never get anywhere. You can actually talk yourself out of real innovation.

"On a small innovation team, you do quick sprints every day and learn on the fly. So part of innovation is just asking again and again, 'What did we get done?' That's why I have every team member send me an update every day. No matter where I am in the world, I get an update, with simple, high-level things. What they set out to do today and what they plan to do tomorrow. That way, they can feel the progress. Part of the progress is being able to have them included in the journey, not just feeling isolated. They need to feel it like a wave. It comes up and down, and it's never perfect. By sending me three or four bullets every day—what I did today, what I'm doing tomorrow—they see short-term goals much easier than feeling overwhelmed by a goal that might seem hard to imagine reaching. Innovation has to have short windows.

"A good friend in Silicon Valley taught me an expression: You have to 'feed the monster a cookie.' Here's what it means. Let's say I've got a big idea that seems far-fetched to you, and you're having trouble seeing how we'll accomplish it. So I start small. You're a monster in my mind. I have to give you a cookie so you'll start liking the idea and not feel so overwhelmed. So you have to break it down into smaller steps. I had to realize that part of having a vision is to be able to translate to people where you want to go and how you'll get there, but with a feed-the-monster-a-cookie approach. People need the buy-in. Give them little nibbles and then they'll get there soon enough."

Fail Fast

You can't have successes without failures, but finding the right balance is key.

"To be an innovative company, it's really important that you don't penalize failure," said Andrew Thompson of Proteus Digital Health. "In an innovative company, and particularly for a start-up company, you have to take risks. So you have to have a very strong bias to action over analytics, and for learning from mistakes and moving forward. That's very much what I call a leadership culture as opposed to a management culture, and it's very counterintuitive to many people who come from large organizations where failure is absolutely clobbered.

"I want to be clear about this: it's not that you reward failure; you don't penalize it. What you focus on much more is risk taking and a bias to action. So the real sin in a small company is not making a mistake, it's not moving. That doesn't always mean you move in the right direction. But if you discover you're moving in the wrong direction, you change direction. It's fairly easy to see and to reward people who have those instincts."

Joseph Jimenez of Novartis talked about the importance of encouraging risk while also minimizing the cost of risk taking. It's a cultural shift that he tried to encourage when he took over.

"The culture in the company, deep down in the organization, has not always been one of risk taking, because it's been quite successful," he observed. "Historically, if you tried something and it didn't work, there was not a lot of reward for even trying. So when I started running the pharmaceutical division, I had to create a situation where people were unlocked from being afraid to fail. And I said, 'Look. There's a way to fail, and there's a way not to fail. And if you can prevent the downside, then fail all you want. I want to see what you're learning.' It was a totally different way of thinking. And there are all kinds of ways to prevent the downside, by being careful about how much you invest. That's not hard to do. But people have to be given the license to try. That's the first step. So it was really about trying to unlock this resistance to try and to maybe fail. And we gave people license to do it by saying, 'Look, as long as you're not creating a huge liability for the company, just go try a bunch of things, and then we can build on them.'"

Jen-Hsun Huang, the CEO of Nvidia, a maker of graphics chips, said he has made risk taking and learning from mistakes part of his company's core values.

"Let me tell you about the two elements that are our core values and that I most treasure and that I spent a lot of time nurturing. One is the tolerance to take risks and the ability to learn from failure. This ability to celebrate failure, if you will, needs to be an important part of any company that's in a rapidly changing world. And the second core value is intellectual honesty—the ability to call a spade a spade, to as quickly as possible recognize that we've made a mistake, that we've gone the wrong way, and that we learn from it and quickly adjust," he said.

"I think 'culture' is a big word for 'corporate character.' It's the personality of the company, and the personality of our company simply says this: If we think something is really worthwhile and we have a great idea, and it's never been done before but we believe in it, it's okay to take a chance. It's okay to try, and if it doesn't work, learn from it, adjust, and keep failing forward. And if you just fail forward all the time—learn, fail, learn, fail, learn, fail— but every single time you're making it better and better, before you know it you're a great company. Mistakes and failures are kind of the negative space around success, right? If we could take enough shots at it, we're going to figure out what success is going to look like. And you have to have intellectual honesty, because you can't have a culture that's willing to tolerate failure because people cling too much to an idea that likely will be bad or isn't working and they feel like their reputation is tied up in it. They can't admit failure. You end up putting too much into a bad idea and then you risk the entire enterprise."

Brent Saunders also encouraged his employees at Bausch & Lomb to cut their losses more quickly on projects that had a low chance of succeeding.

"R&D sometimes can be the black hole of spending in health-care companies," he said, "and so we wanted to also create incentives for scientists not to chase dreams that had low probability of success. So one of the things we did was to make sure we celebrated and rewarded scientists for killing things early, too. Because a lot of times in life sciences or health-care companies, these projects become like a child to the scientist. It's what they work on. They fear that if they get rid of it, there won't be a need for them in the organization and so they continue to spend and figure out ways to keep their project alive against very low odds.

"Culturally, I think the way you crack that is by saying that you should celebrate your successes and that you should equally celebrate your fast kills. We all fail at things. It's about failing intelligently and failing fast so that you don't waste money chasing something that's never going to make it out the door and into customers' hands."

Symbols can also help reward risk taking. That's one approach used by Kyle Zimmer of First Book.

"Occasionally, we give out a Brick Wall Award for an idea that should have gone really well, but ended up crashing into a brick wall," she said. "It's a way of saying, 'It's okay, you did the thinking, and you gave it your best shot, and it crashed, but it was an honorable step.'"

Crowd-Source Online

Corporate intranets have made it easier to harness the collective wisdom of employees to come up with, refine, and ultimately choose ideas to invest in. John Donovan of AT&T describes the process he used to spur innovation at the telecommunications company.

"I started by going to my direct reports and telling them I wanted the sixteen smartest people in the technology organization," he said. "It's not about titles. I don't want any diplomats. I don't want any process people. And I called them the Tech Council. I rotate people in and out. I gave them several hours a month. Rules of the road are that you're not allowed to report back to anybody you work for—what is said in this room stays in this room.

"We then started with a list of all the things that were broken, stupid, idiotic, what's killing innovation, from sixteen really bright people who were willing and able to tell you the truth. And if you look at some of the things that we've done in our innovation program, a lot of the seeds were born in that room. And so we built a profile that started with the ugly truth, and that's kind of where we had to start from. When I came in, I was led to believe that we would have an innovation problem. And I learned very quickly that we did have an innovation problem, but we didn't have an invention problem—and that's an important distinction. If you're going to build a car and you have no blueprint, and you have no factory, that's a different place to start than if you go in your garage and you turn the lights on and there's everything that you need to build a Ferrari. It just isn't assembled. That's a very different world. So I very quickly started to tell people that we're close. We have inventions. Most companies don't have that. I describe innovation as invention with a customer at the other end. When it becomes a relationship that's consummated with a commercial transaction, then that's an innovation. That adoption part of it was where we were not good. But the invention was there.

"So I did a diagnosis. The sixteen folks who were with me gave me a million reasons to explain why we couldn't innovate, it's just simply impossible. So then we spent hours and hours just breaking the problem down. Then we said, 'Let's start an innovation program in this room.' And we started our 'crowd-sourcing' plan in that room. It starts with the fact that we're too kind as an organization, so no one stands up and says, 'Your ideas are lousy.' So we have this culture of really nice people, and it's nurturing. And there's a convenient place to blame if I don't want to tell you your idea isn't any good—I just blame it on the people upstairs, and I say that they don't like it.

"So I said, 'Let's start with some tabletop exercises.' Each of us brought six ideas into the room. So we had ninety-six ideas. Everybody presents their ideas, and then we voted on them, and chose the top two that were going to get funded. And then we knocked down every hurdle to get these two ideas into their final form and into a funding process. Today, that idea has been expanded to an online site, with more than a hundred thousand people who are on the innovation pipeline, as we call it. It generates thousands of ideas. Then people vote on them, like in social media—thumbs up, thumbs down—and it's invisible who voted. Why? Because, culturally, people don't want to call other people out. So then you move to the collaboration phase, where you need comments, and comments are not without attribution. We thought about it, and decided that you have to be accountable when you use words. And so it doesn't feel like a big company anymore. Somebody might say that an idea is impractical because it would be really expensive, but then someone says, 'Wait a minute, what about trying this?' So the group comes together and they solve the problem.

"If you look at this process, it's designed around the idea of venture capital, because I had the experience coming out of that world on how things get funded. It also uses social media to solve this perceived problem that there's some corporate machinery that kills ideas. Now it's your peers, and good ideas ultimately get funded with real money."

For employees at Red Hat, which develops open-source technology as its business, using a crowd-sourcing model for innovation was a natural step. Jim Whitehurst, the CEO, explained how the company uses a program called Memo List to share and sharpen ideas:

"Since we were founded in the 1990s on the idea of leveraging broad open-source communities, we naturally adopted that approach in our culture long before the Facebooks of the world even existed.

So we're on the bleeding edge of what so many companies are going to face because of this whole millennial generation coming up. It just does not like this idea of hierarchy.

"If you really want innovation to happen, you almost have to think about it as an ecosystem. A lot of companies think that the way to be more innovative is to put a group of creative people together. But your most creative ideas are going to come from people on the front lines who see a different way of doing the jobs they do every day. You have to create vehicles for those ideas to be heard. So the question is, How do you make that happen? How do you engage your employees? And what we do is to use social media internally.

"A lot of people think that using social media for engagement means you're somehow creating a democracy, or at least consensus. But we have a culture of meritocracy, not democracy. And the difference between meritocracy and democracy is night and day. We let debate happen, and you let it kind of burn its way out, with people offering their opinions on both sides of an issue. And then you say: 'We've listened to all of this. We've taken it into consideration and here's what we're going to do.' Even the most ardent people opposing whatever decision is ultimately made will at least think, 'I had my say. You heard me, and you told me why you made the decision.' It does not have to be a democracy. And this has been true at Red Hat since long before I got there.

"A lot of the issues that many companies are now facing are because they think, 'I can't let my employees have a seat at the table in this.' But it's not about having a seat at the table for the decision. It's about having a seat at the table to voice their opinions and make sure those opinions are heard. As long as our employees are involved, they will accept virtually any decision. They may say, 'We don't like it, and we still don't agree with that.' But you listen and you come back with a well-reasoned answer. And that is the expectation that our employees have. I think almost every company is going to have to deal with this over the next twenty years.

"Our internal forum is called Memo List, and on average you'll see a couple hundred posts a day. And I go through it every single day. I would say probably three-quarters of the people are on it every day, either reading or posting. It can be time consuming, but here's how I think about it: Engaging people in how decisions are getting made means it can take forever to get decisions made. But once you make a decision, you get flawless execution because everybody's engaged. They know what you're doing and they know why you're doing it. It's a very different model than what happens in most companies, which is that a small group of senior people make decisions, and then execution is difficult. I'm not saying we're perfect by any stretch, but by the time we've actually come to an agreement on where we're heading, we're halfway there."

Daniel T. Hendrix, the CEO of the flooring company Interface, said his company has developed an online innovation platform that also helps bring in ideas from people outside the firm.

"You have to keep reinventing yourself in this world," he said. "Everybody is a fast follower and so you've got to find a way to create an innovation environment, and it's all around collaboration. It's all around engaging the shop floor all the way up to the top. You really have to fund, fuel, encourage collaboration and engagement to get innovation and you've got to go inside and outside to do that.

"We developed something called the Innovation Farm. It's a platform where everybody can participate within the organization, and you can also go outside with it. So you pose questions. 'How would you solve this problem?' And everybody gets engaged in how you might solve that problem. Or you just ask an open-ended question. 'What do we need, from an innovation standpoint?'

Then people vote, and the best ideas surface over time as people vote and start tweeting and talking about that idea. It really got our people engaged in this whole open-architecture structure.

"And we really tried to get out of the box. We didn't want a closed architecture because I think we had a closed architecture to some extent. We really wanted to create an open-architecture structure, for anybody in the company, and even outside. So we're going to post questions on the Web. 'How would you solve this problem?' And then if you get something that's really interesting, then you just close off the conversation and discuss it further in a smaller forum."

Crowd-sourcing innovation does not necessarily require elaborate intranet platforms. Vineet Nayar, the CEO of HCL Technologies, a consulting firm, said he will simply toss out ideas and problems on his internal blog, and employees provide their input.

"I used to write a blog every week because I thought people wanted to know what was going through my head. But one employee told me that 'Actually, we want to participate in solving a problem.' So, the blog got converted into me asking a question: 'This is a problem I'm having. How will you solve it?' This is one example of how we started going in one direction, and the direction completely changed to another."

Crowd-Source Offline

Harnessing the insights and opinions of all employees doesn't necessarily require technology. It can simply be a matter of some tape or pushpins to put work on the walls, and then to encourage everyone to feel that they can share their opinions to make the work better. Doing so requires a culture where critiques aren't necessarily taken as criticisms.

"Collaboration is a word that's overused a lot, but we do practice it," said Christine Fruechte, the CEO of Colle + McVoy, an ad agency. "Our offices are completely open, and all of our work is posted on the wall. So if you've got a big ego, leave it on the elevator, because the creative business is a very vulnerable business. We put up all of our ideas, whether they are strategic, digital, media, and they are there for everybody to see and to give feedback on.

"You have to have a lot of confidence in the notion that the endgame is the best idea. It's not about whether you look good, or are the smartest person in the room. Every idea, every project that we're working on is basically all on the open wall. So if I'm gone for a week, I can come back and literally walk along the walls and catch up on the majority of everything we're doing. People will rewrite headlines. People will say we need ideas for this or that. We post it all up there. And then you don't know whose ideas they are. You just start circling the ideas that you think achieve the objectives."

Mark Fuller of WET Design said he also likes to have his employees put their work on the walls for others to see.

"If you've got a drawing, you can stick a couple of magnets on it. The point is to get people to put their stuff out where other people can see it," he said. "We don't want a culture of 'That's my idea. I don't want anybody to see it. Maybe they'll find a flaw in it.' I had a teacher once who said, 'Whenever you guys are sitting here, and you realize that you've made a mistake on something you're working with, I want you to applaud yourself.' He said, 'That will accomplish a couple of things. First of all, instead of saying, "Oh, I made a mistake, I'm never going to learn this stuff anyway," you're going to reward yourself because you caught the mistake before I did.' We all rolled our eyes in the class, but I've never forgotten that. So one of the things I will do is to start some meetings by saying, 'Let me tell you where I just screwed up.' That sets the tone of 'We've got to put our mistakes out there.' They don't call it 'learn by trial and success.' You learn by trial and error."

Go Smaller

Numbers seem to work against innovation. The more people attend a brainstorming session, the more people are assigned to a project, the more likely it will slow down. John Nottingham and John Spirk, the co-presidents of the innovation consulting firm Nottingham Spirk, see the dynamic every day.

"I'll tell you there's nothing more frustrating," Spirk said, "than to sit in on a conference call, and we'll have maybe three people on our side, and there will be fifteen or sixteen people on the other side of the call. The bigger the company, the more people tend to be on the call. And once we kind of figure that out, we have a talk with the company because that is no way to innovate. It just does not happen with sixteen people on a weekly conference call.

"We'll tell them that we need to narrow the team down, and get the key people involved, the ones who own the project. Not just people who are watching it, but the ones who own it. And there's usually only two or three who truly own it."

Nottingham added this perspective: "The analogy we use is that you're in a Jeep SUV on one side of a pond of quicksand, and way on the other side is the place you want to get to. That's the goal. And you rev up your engine and start moving and you're negotiating this tricky pond. If you pause to think about it for a while, you start sinking in the quicksand and all of a sudden you disappear.

"The more people you have in a meeting, the more you have the opportunity for a naysayer—'Wait a minute, what if it doesn't work? What if it doesn't happen?' All of a sudden the dynamic just slows down. The right size for an effective brainstorming meeting is between eight and twelve. If you get too many people in the room, it's not collaborative enough. And if it's too few, there are just not enough good ideas. But we've been to Fortune 50 companies where they had a creative session where there were about a hundred people in the room. It gets unmanageable. And what happens is people end up shutting down."

Lily Kanter of Serena & Lily said that start-ups have to make sure that the need for process doesn't overwhelm the entrepreneurial spark of an organization; small teams can be an effective way to guard against this dynamic.

"You always need entrepreneurial people who can wear a hundred hats forever," she said. "It's just finding the right positions for them, because they're fantastic. They make stuff happen. And then you bring in process-oriented people who have these deep backgrounds in a particular area of the business. You need that balance of both those types to build your organization. But you have to be careful, because the more process you get, the slower and more bureaucratic and less nimble you get. Things can get bogged down. You want to start a new initiative, and then get fifteen people in the room to discuss it. I think keeping that entrepreneurship ability alive in a company is really important.

"So we go rogue. We tell people that if there's an innovative project we're working on, it's going to be a small team. I think you need to be disruptive. And I think entrepreneurial people can be very troublesome to certain organizations if they're constantly disrupting processes. Of course, you need process. You need calendars. You need people knowing exactly what they're doing every day to deliver the plan. So it's important that when you do have disruptive, innovative projects, that they're literally not happening inside the organization, so that the current state of the organization is not turned upside down because of it.

"So ideally you have no more than four on a project team. It really depends on the project, but somebody's just got to own it. Two to four people on an innovative project is more than plenty. After all, how many companies get started with more than that? I think small working groups are very efficient. It's about making sure people understand where decisions lie."

Jen Guzman, the CEO of the pet food company Stella & Chewy's, uses a similar approach, breaking down larger brainstorming sessions into smaller groups.

"The typical meeting we have is about four people—the people who are really going to be actively participating and giving ideas, versus a much broader group. I just find that for a small company that needs to be very nimble, that's actually a good way to operate," she said.

"But there are other meetings where it's more of a brainstorming session, where it can be a very large group. During one recent one, we probably had about twelve to fifteen people. Basically I communicated, at a very high level, what we were trying to achieve, and not much else, because I didn't want to sort of feed them the ideas. I wanted to hear what they had to say. In that situation, we broke down into four different groups, with about three to four people each, and had them go into different rooms to come up with their own ideas. And then we regrouped, and had every group present, and took the best ideas from each group.

"I think it's very difficult—it can almost be unmanageable—to have a brainstorming session with fifteen people and get complete ideas from it. This way, you're getting everyone's input, because by breaking them down into smaller groups, you have several meaningful discussions happening, and then when everyone regroups, you have yet another meaningful discussion about the best of everything. I like to be able to follow thoughts through. Very big meetings can become a bit less thoughtful sometimes and less of a dialogue. When you have a large group, and you just sort of put them all on the spot, you get bursts of ideas, but you don't get as much thought."

The keep-it-small approach can be applied not only to meetings, but also to divisions within a company. Andy Mills of Medline said that his firm continually creates new divisions so they can move quickly.

"Creating smaller divisions actually works pretty well," he said. "We take areas that we think are starting to get too large. So there may be part of a larger division that is not getting a lot of attention. So we break it off and give it its own identity. We try to work in business units that are probably smaller than at a lot of other companies, and that's why we have twenty-one product divisions and they're fairly autonomous. We'll theoretically add a bunch more as we grow. A lot of it is based on gut feel about the bandwidth of the supervisor or manager of each division. Are they able to give enough attention to everything in the division they're running?"

Try This Nifty Card Trick

Many meetings have a predictable dynamic. Ideas are presented, and people are asked for their opinions, but everyone is focused on the most senior person in the room. If the boss says he likes the second of the five options on the table the most, then guess what? It's pretty likely that a lot of other people around the table will agree that the second option is indeed—no question, hands down, by far—the best.

John Nottingham of Nottingham Spirk recommended a strategy for guarding against the gee-boss-you're-right problem:

"There is a technique we use and it's very simple. The first phase is to generate a lot of ideas. People often do that pretty well. The frustration that I've seen is when people say, 'We had that creative session six weeks ago. Whatever happened to that?' There's a whole notebook full of ideas, but they're not implemented.

"For the second phase, let's say you have a hundred ideas, but you're only going to do one or two products. And that's the part that most people don't get. One technique we use involves notecards. Everybody gets three three-by-five notecards, and one note-card says, 'Wow.' One notecard says, 'Nice.' And the third notecard says, 'Who cares?' Everybody sits around the table with their cards facedown.

"Now you have somebody champion one of the ideas on the wall and talk about why it's a good idea. When that is done, everybody—at the same time—has to hold up a card, and then you look around the room. Now if after that presentation everybody says, 'Wow,' well, wow is an emotional word, and you're going to keep that idea. That's easy, but it doesn't happen a lot. So another idea is presented, and everybody may hold up a card that says, 'Who cares?' So you take that product and just shove it off the table. It's not going to work.

"Guess what the hardest one is? 'Nice.' So the product manager comes in, talks about this thing, and everybody says, 'Yeah, it's nice.' You know what? Nice is the hardest thing because somebody will actually produce the nice product and will have nice sales but it's not going to move the needle. Too often, too many nice products get produced. In these meetings, it's very important that everybody shows a card at the same time so nobody influences anybody else. You don't go around the room taking turns, as in, 'What do you think, Harry? What do you think, George?' Well if Harry's the leader, then everybody starts answering the same way."

Hold a Hackathon

To get creative juices flowing, many leaders will hold variations of "hackathons," in which they set aside time to focus intensely on creating, brainstorming, and developing new ideas. There's no particular formula beyond making sure people get out of their normal routines. Google famously lets engineers use 20 percent of their time to do company-related work that interests them personally. Andrew Thompson of Proteus Digital Health has adopted a clever idea that is making the rounds in many organizations: "FedEx Days." The name comes from the delivery company's promise to deliver things overnight.

"Pretty much anyone can apply for FedEx Day, or any group of people," Thompson said. "The deal is that you can take the whole day and go off and do something, but it's FedEx, right? So it has to absolutely, positively be delivered overnight. And you can break it down, because maybe you want four FedEx Days, but there has to be a deliverable for every day. That's also terrific because it's everything from very simple little things—like improving the layout of the desks in the area—all the way through to fairly important things."

Some CEOs even believe that upending employees' normal sleep patterns can lead to some good ideas. Aaron Levie, the CEO of the online data storage company Box, described the simple structure of his hackathon:

"The engineering team pulled an all-nighter, from eight p.m. until noon the next day, on projects outside their daily job description. We then had a judging panel at lunch, and the entire company got to watch the engineers present some amazing new features. It was fun and people goofed off but it was also really inspiring, and I think it brought the whole group together."

Jarrod Moses of United Entertainment Group said that he uses a tour bus and "flip days" to generate new ideas.

"One of the first things we bought was a tour bus. We use it instead of flying. We take it at least twenty-five times a year to different meetings throughout the country. There's an amazing culture that develops on the bus. You learn so much about one another, and you develop this candor and trust that you don't get in the office. The creative juices just flow, and they flow twenty-four hours. They could come from a joke; they could come because someone is just overtired. You never know. The point is that there's no barrier to entry for the idea. People are wearing T-shirts and shorts. No one is the CEO on the bus. It's like a band. There's a magic to it.

"A few times throughout the year, we also flip the day for the staff. You come in at four in the afternoon and work until about four or five in the morning. You mess with somebody's internal clock, and some interesting ideas come out. It kind of clears your head."

Michael Lebowitz of Big Spaceship set aside time for people to just work on projects that are not necessarily for a specific client.

"We do a lot of work just for ourselves, creating products, pieces of content. I invest fairly heavily in that, in having time," he said. "We did a thing one summer called IP [intellectual property] Fridays. You take the traditional corporate summer Friday where everybody's supposed to be allowed to leave at two p.m., but everybody has to work anyway so they can't, and they just feel miffed. So we get a big lunch for everybody and at two p.m. on Friday, we close to client work and spend from two to seven working on our own internal projects. And the ideas for those come from anywhere in the company.

"We have a little form with a few simple questions on our internal blog, and then a few of us vet the ideas. We want them to be simple, because we want small things that we can act on quickly. So we've got all this stuff out in the world that we created for ourselves, and people get excited about that."

Keep Entrepreneurs In-House

How do you tap into the entrepreneurial spirit many employees have, so that when they come up with a big idea, they feel they can run with it, rather than leaving the company? Here's the system that Robert LoCascio put in place at LivePerson:

"Over the last two years, we created four or five new products out of nowhere. We created a whole new product innovation structure, where anyone can create in the company now. The new products came from people who you wouldn't actually expect to think of new products, but they were close to customers. They thought up a product, and then we said we're not going to have a formal process, but you're going to be an entrepreneur internally. That means you're going to beg, borrow, and steal. You're going to get customers up on this thing. You're going to get people to believe in you, and then when you get a couple of customers and the product is working for them, you come back to the leadership team, which is about twenty-five people, and present us your business plan and we'll fund it. So we funded these business plans, and then these products went to market. We've gone from a one-product company to a five-product company in under a year, and now fifteen percent of our bookings are from new products.

"And we have a compensation program where we will value their business five years from now and pay them out like an entrepreneur. Because what I found is that these guys normally leave and they get funded somewhere. So I said, 'Stay with us. We'll look at your revenue stream and we'll put a multiple on it and we'll give you ten percent of that.' And so there's a payout for them to stay."

Creating these start-ups within companies is essential to holding on to the entrepreneurial culture as firms evolve from start-ups to mature companies, said Jeremy Allaire of Brightcove.

"You always need to have everyone feel like they're on some broader mission," he explained. "In the early stages of a company, the mission is: 'Are we going to survive? Is there a product? Does it work? Is anyone going to want it? Is there a market?' They're like existential questions for a business, but I think those core mission tenets remain important through that growth stage. It's something that people can attach themselves to, so people aren't just coming in to a job. So I've tried to really provide that narrative over and over, even as the milestones just keep changing.

"The other thing is to hold on to that feeling of being a start-up, and it actually relates very deeply to how you go to the next phase of growth. Companies that figure out how to really become significant in scale reinvent themselves and create completely new things. Just as an example, a little over a year ago, we created a start-up inside the company to create new products. That was so galvanizing and so energizing, and it kind of cascaded across the company. People were saying, 'This isn't the kind of same old, same old—we're reinventing ourselves.'"

Lynn Blodgett, the CEO of ACS, a computer services company that was acquired by Xerox in 2010, developed a philosophy from the earliest days of the company to continually drive down accountability, and develop the right incentives, so as to tap into the entrepreneurial spirit of employees.

"I believe that a really important management principle is that if you get the incentives aligned, people will motivate themselves far better than you'll ever motivate them," he said. "But, again, you have to get the incentives right. It's not only financial. It's being able to feel like they have a level of control over their destiny, that they are valued in what they do, that they're being successful, that they're contributing. Those things are actually probably more important than the money. But you've got to get the money right, too.

"Getting everybody's interests aligned on the same side of the table, including the client's, is key. Because that's not normally how it is. You have the clients with their objectives. You have the employees with their objectives. And then you're over here trying to make nice with everybody. I think that the more direct the accountability, the greater the performance.

"I remember how it was when I was starting out. We had an investor, and we were just teeny. We were doing about thirty thousand dollars a month or so in business. ACS is almost eight billion in annual revenue, so it's a little more complex. But you know what? The principle is the same. When I was responsible for that little business, I had my bar charts, and I'd color them in every day. That's how I kept track, and I knew that if I didn't make my number, it was my responsibility. Well, what we've done is we've said: 'Okay, how can we create that entrepreneurial drive deep into a large organization? How do you do that?'

"One of the ways that we do it is we drive the P&L as deep into the organization as we can. We have a P&L at a customer level, that's mandatory. We have to be able to see how we're doing with that customer. A lot of companies can't do that. In our business, we drive the P&L down to the people who are actually doing the work. So if we can make a P&L for a ten-million-dollar business, we'll give that guy the P&L and he'll have profit accountability, revenue accountability, and customer satisfaction accountability. And as they grow, they make more money. That results in a higher performance, in my opinion.

"So you give people control, hold them accountable, give them control of their resources, and then monitor what they do. And if you do that, you're going to tap into, in our business, the highest level of drive—entrepreneurial drive. I want ACS to look like a whole bunch of sole proprietorships. Because that way, people are thinking to themselves, 'If this was my money, if I was doing this, would I really spend it? Do I have to buy that computer right now or can I get by with one that's two years old?'"

Competition among divisions for resources can also drive innovation. Andy Mills explained the approach at Medline:

"I like to refer to it as Darwinian, but it's also a very democratic approach. One of the biggest assets of the company is our sales force. They're approaching thirteen hundred people now, and the philosophy of the company has been that we should let the sales team and senior sales management decide the emphasis on which products we should be pushing and promoting. We sell three hundred and fifty thousand products, but we promote certain key products or larger product lines throughout the year, and it's not senior management who's choosing them. We bring in ten or twelve sales reps, part of a kind of roving committee, and we bring in a few senior sales managers. Then we have twenty-one product divisions present, and they basically have to explain why they should get the extra time and training and extra sales emphasis from our team.

"I'll have one vote, but the committee basically picks, and we've coined it as a 'survival of the fittest' approach, because the product divisions are competing against each other for the equivalent of shelf space in our world. It's really become part of our culture and really a good thing because when somebody isn't selected, they get an explanation, and they're told to go back to the drawing board. And then next quarter, when they present, hopefully those issues are taken care of. So we are constantly raising the bar.

"We developed this approach because, when the company was started by my father and uncle, they were not only sales reps themselves, but they did other things too. It was a small business and they just wore many hats. So they thought it was equitable to share the decision making with others who were sales reps like them. We've always felt that somebody who's out talking to customers has better insight than somebody who's in a marketing or product division. We do a lot of listening to the market, and for us the market is both our customers and our sales reps."

Reverse the Peter Principle

The traditional approach is to take high performers—people who truly excel at a given task—and then move them into management. The thinking, of course, is that they will help elevate a broader group of people to be as good as they are at that work. It makes sense in theory, but often doesn't work in practice, and the company loses the spark it got from letting those stars simply do the work they do best.

Tracy Dolgin, the CEO of the YES Network, has adopted a "Reverse Peter Principle" as an antidote to this dynamic.

"Let's start with the Peter Principle," he said. "Most people start out as doers, and they have a function—they're a marketing person, a human resource person, a finance person, a production person. And they get really good at doing that as they gain more experience. The reason they usually get promoted is not because someone innately thought that the person would make a great manager. They get promoted because they were a great doer. Is the same person going to be a great manager? Sometimes yes, sometimes no. The Peter Principle holds that as people get promoted again and again, they keep doing less and managing more until they get to a point where they stop getting promoted because they're not as good at their new job. The Peter Principle says that you end up in the job you're the worst at.

"So I said, let's try the Reverse Peter Principle. The company was small enough so that if I found the best doers of every single critical function and convinced them to come to the YES Network to manage less and do more, to create a flat organization where they were going to be the best doers in the world, we would actually be able to create something incredible. That would allow us to compete with the big boys and maximize the business by out-executing them. Just change the rules and you could compete with them, because now I could out-execute them because my executors would be the best people at executing, and they would be spending 90 percent of their time executing and not managing.

"So when I went out and recruited, I basically told people my theory, and that we were all, including myself, going to go from being a manager to a doer. I was looking for the best people in each particular function. Did they have the passion? Did they believe that we could win that way? Those were some of the questions I asked. Did they think that this was a good idea or a bad idea? Before I even explained this to them, the first question I asked was: 'When you're working, what do you love doing?' I looked for people who would talk to me about doing, not managing.

"I also looked for people who've done everything. They've had long careers. And they said, 'Frankly, I'm just ready to roll my sleeves up again. I'm tired of spending ninety percent of my day dealing with management, which is usually dealing with people's problems, not even business problems, and the politics. If I could just spend my energy on doing this . . .' So I had to find people who were the best doers, who were willing to give up the management part of it and thought this could work. We wanted it to be more like a start-up environment, except we didn't have a start-up product.

"The Reverse Peter Principle is also a fantastic thing because I can pay you like a manager to be an executor because I've got far fewer people: no layers, and a flat organization. The best doer who has a lot of experience can do more things than five or six or seven new doers. And so we didn't have to build up a big company. I could pay a small group of people really well, like the big media companies pay them, to be doers.

"There have been people who have absolutely failed in this system. You've heard this a million times before: The only way to fail is to not fail, because otherwise you're not taking risks. You're not getting better. You're just doing the things that you know will work. Now, the difference is that you really want people who learn from their mistakes. There is not much of a benefit for making the same mistake two or three or four times."

I asked Dolgin whether any lessons about the Reverse Peter Principle were transferable to other companies.

"Every company of a certain size has a lesser version of a Steve Jobs in every function," he said. "How do you find those people and make sure that they're not getting burned out trying to be a doer as well as a manager and a politician? How do you create a kind of skunkworks—a place inside of a big company where you foster innovation, and where you reward doers for that task? You can't run the whole company that way, but there are lessons to learn that I think could really help. Sometimes the best people, no matter how many layers up, have to become doers again for mission-critical tasks. Some companies have twenty layers—they call them pyramids for a reason.

"I think that at every layer of that pyramid, you've got to be able to take some people who are just so extraordinary at what they did before they became managers—the doer part of their job—and free them up to spend maybe fifty percent of their time doing as opposed to managing. That's the lesson that you could take to a big company."

Other companies have adopted this philosophy, too.

"I like to hire confident people, people who are comfortable not sticking to a title or an organizational structure," said Selina Lo, the CEO of Ruckus Wireless. "I want people who are comfortable with their staff being paid more than they are. That was one policy I was clear about early on—that in our company, managers should never assume that the people who work for them are paid less than they are, or get fewer stock options. In fact, this is something that I have to continue to reinforce because we have new managers.

"This approach came from my time working at Hewlett Packard. If I don't have the patience to be a manager, and yet I make a big contribution to the company, why should I be limited by a conventional career track that says that I have to learn to manage people in order to get there? So in my company, there is a rule that all new managers need to know—that it's not a given that their people will be paid less than they are. That's part of becoming a manager: that you really have to enjoy enabling people, more than doing the actual work. And so I want people who are good managers to be managers. I don't want people to become managers just because they feel they need to.

"I also don't want the company to be too process-oriented and to be too hierarchical. You don't want processes and structures to become boundaries for how to execute. If somebody can help out with a project, even though it's not part of their responsibilities, I have no problem going straight to that person to get them involved. Sometimes, it really irks my staff, but they understand. And as a start-up, you will never have enough resources to staff every function, so you're going to have to keep that boundary a little gray.

"I've also found that people who are entrepreneurial like to not be defined. They really like to go beyond what they are told to do, and they would put in their own time, and put their heart in it. If you give them the opportunity to extend, employees like it. I like it. I just need the people in the middle to make sure that it doesn't go out of control."

Allow "a Little Bit of Chaos"

Part of the job of leaders is to reduce chaos, to create structures that cut down on inefficiencies and that streamline processes and work flow. It can be counterintuitive, then, to loosen up to allow for some chaos. But that is what many leaders said they've come to understand is a necessary part of developing an innovative culture.

"This is an art form, but you have to understand that the more diverse people are, and the more uncomfortable they make you, the more likely you are to create something new together," said Charlotte Beers, the former chairwoman and CEO of Ogilvy & Mather Worldwide. "Entertaining a certain amount of discomfort is a very important management tool. In the digital age, people put too much stock in logic, data, the facts. But all the excitement and innovation is in the illogical, and the interpretation of this data by someone who's still breaking the rules. As a manager you have to be able to entertain chaos. Otherwise, you'll never get close to the creative process. I worry that more and more today we lean toward that which is provable, when all the excitement is in the hard-to-prove."

Michael Lebowitz of Big Spaceship said he provides a budget to encourage a bit of chaos.

"I think a little bit of chaos is really good, as long as people can focus on what they need to do," he said. "In some ways, I just try to enable stuff to happen, and I'm not quite sure what the stuff will be. We give each discipline a monthly budget, a sort of use-it-or-lose-it budget. Spend it on whatever you want, as long as it's related to stuff you're interested in. Designers will buy craft supplies because they want to do stuff with paper and then shoot it and composite it on the computer. Strategists buy books and go to conferences and things like that.

"This is not about creating a utopia. I mean, we're a profit-driven company. But I want a little bit of chaos to be a natural part of the company. Even though that makes things sort of harder, it doesn't necessarily make them less efficient. I think that's an assumption people make that's derived from an industrial era, when efficiency was the defining thing. But it's not defining for what we do. And what we find is that it's more efficient for us to do things that would seem less efficient in another industry."

For leaders, it can take time to grow comfortable with a certain amount of chaos—a lesson that Marjorie Kaplan of the Animal Planet and Science cable networks said she had learned over time.

"I think organizations have a tendency to be self-censoring, always moving forward and making decisions at the exclusion of a tolerance for confusion that I think you need for creativity. You don't want a confused organization all the time. But you can't have an orderly organization all the time, either. So I think real creativity comes from the ability to tolerate the confusion and to be able, in the right moment, to land the decision and then move forward with that decision," she said.

"I think my ability to tolerate confusion has grown. That does present management challenges because not everybody can tolerate it. And organizations can't be confused. I've come to rely more in different ways than I have in the past on my senior team in those other areas. Some people are really good with order and organization, and I rely on them to have that function and to support me in that. And I'm overt about that.

"Understanding the right amount of chaos comes with experience. I mean, other people could tell you whether I'm good enough at it. But I think that creativity is frightening and messy. You're not going to get a big, game-changing idea from trying to do what you're doing now, but just a little bit better. And so you must find a way to try ideas that don't seem like they make any sense, to let people who are those kinds of people just go spiraling off for a while, because that's their process. And then decide when it's the time to reel them in."

Look Ahead, Look Back

Two CEOs offered smart step-back thoughts about discipline: the discipline to stay focused on the work that has already been launched, and the discipline to always look ahead to ensure that companies don't become complacent.

Geoffrey Canada of Harlem Children's Zone argued that there is a shelf life for innovation.

"Innovation sticks for about eighteen months," he said. "So let's say you put a great innovative program in place. You put the right people on it, you get everything organized, and then if you don't come back and do anything with it for eighteen months, that program's half as good as when you started it. They just start decaying.

"And I think one of the challenges for us in this business, in management generally, is that nobody wants to keep going back and doing the same thing over and over. Everybody wants to get this brand-new idea and really get it going, instead of paying attention to the other things that are fundamental to our business. If you don't go back and check on a regular basis, those things begin to decay, and you end up constantly having to reinvent something that you already did. Getting a team of people who really understand how essential that is to staying great is one of the real challenges."

Carl Bass of Autodesk described his thoughts about how to continually foster innovation:

"This is my current fascination: it's this whole idea about keeping companies entrepreneurial and innovative and cutting edge. The thing that I worry about a lot is how companies measure themselves. The analogy is that you can see light from a star that burned out a long time ago—it's a hundred light-years away, and three years ago that star died.

"The same thing is true in companies. We measure ourselves around revenue and profits and financial metrics that perform long after a spark is gone. You have this funny feedback mechanism in which you're getting the results from something that happened a while ago. Maybe the thing that generates all the revenue was a great idea that happened in a dorm room. There's a lot of stuff that's gone on since then, but do you know whether you've had another spark?

"So a lot of my time these days, as I've gotten a little bit older and more reflective about how you manage a big organization, has been about trying to identify whether real things are going on. I've been spending a lot more time trying to quantify or figure out if what we're doing is right, or whether what we're really doing is just celebrating the result of things that happened a while ago. I think it's real easy as a leader to confuse what the results are today with the actions that happened a while ago, because then you just start coasting."

Count to Twenty-Four

A final thought on innovation: any leader who wants to encourage it would be well served to remember the "rule of twenty-four," explained here by Tony Tjan, the CEO of the venture capital firm Cue Ball.

"One of my mentors was Jay Chiat, one of the founders of the Chiat/Day ad agency," Tjan said. "He had this incredible capacity for optimism, particularly optimism during mentorship. He had this amazing ability to think of every reason why an idea might work before criticizing it and thinking why it might not work. When you're a mentor, you've got to realize that people are often sharing their dreams, and I think it's human nature to be a critic. We're skeptics. As you get older and more experienced, wisdom is great, but you also have to be careful not to automatically impose your mental framework and your lessons.

"I've translated it into a rule that I try to get people to follow, and I'm still working on this. When someone gives you an idea, try to wait just twenty-four seconds before criticizing it. If you can do that, wait twenty-four minutes. Then if you become a Zen master of optimism, you could wait a day, and spend that time thinking about why something actually might work. In venture capital, you're at the intersection of human capital and their big ideas, their dreams. My favorite quote is from Eleanor Roosevelt: 'The future belongs to those who believe in the beauty of their dreams.'"

15.

CAN WE HAVE SOME FUN?

Humor, to me, is the world's best tonic.

—ANDY LANSING,
CEO of Levy Restaurants

Few people have mined the dark side of corporate culture as profitably as Ricky Gervais and Steve Carell, who played the office manager in the British and American versions of the television comedy *The Office*. Week after week, the shows delivered cringeworthy moments. "I swore to myself that if I ever got to walk around the room as manager," said Carell's character, Michael Scott, in one episode, "people would laugh when they saw me coming, and would applaud as I walked away."

In a 2009 interview, Carell said of Scott: "I think he's a man who clearly lacks self-awareness. And I've always said that if he even caught a glimpse of who he really is, his head would explode. As Ricky said about his character, and I think it applies to Michael Scott too, 'If you don't know a Michael Scott, you are Michael Scott.'"

Even at a subconscious level, this fear—of someone muttering to a colleague that they feel like they're living in an episode of *The*

Office—probably makes a lot of leaders reluctant to try anything fun at the office, worried that some idea might bomb, leaving them like a stand-up comic on the stage, hearing only the sound of crickets. The safer course for managers is just to keep everyone focused on the work.

Too bad, because there's nothing like some good, honest fun and a few shared laughs to bring people together and provide some glue for the team. As the musical humorist Victor Borge said, "Laughter is the shortest distance between two people."

So what are the secrets to having fun at work?

At a certain level, trying to answer the question is like trying to break down a joke: you can try to parse why something is funny, but ultimately humor comes down to timing and delivery. Some companies have pajama days or Disco Fridays (quick breaks when everybody dances in the hallway), and Halloween certainly seems to be the High Holy Day of American business, given the number of CEOs who told me how fully their employees embrace dressing up and decorating the office.

Why do these events work at some companies? It's hard to say, but one obvious strategy for avoiding the Michael Scott problem is simply to let employees come up with the morale-building ideas themselves. They might need, say, time and a budget for outfitting the office with Halloween decorations. The leader's job is to go along for the ride, not to play camp counselor.

Let the Boss Be the Joke

One tradition at First Book is Dinosaur Day, when the staff dresses up in T-shirts with dinosaurs on them. The point, explains CEO Kyle Zimmer, is to make fun of her.

"It started because when I explain what First Book is, I say as an example, 'So if you are a teacher—if you're Mrs. McGillicuddy—

and you want to teach your third graders about dinosaurs, and say March is Dinosaur Month . . .' What I didn't realize, of course, is that I had said that same thing fifty thousand times. So now there is an annual Dinosaur Day event, to celebrate the fact that I'm not very creative in my pitches, basically. We have a big lunch, and people serve collective humiliation. It's very funny. And somebody always comes in dressed as Mrs. McGillicuddy."

Of course, the CEO has to be able to laugh at herself, and employees have to know, at a gut level, that the boss will enjoy the send-up. Chris Barbin of Appirio explained in chapter 3 that having fun was one of the company's three core values. One clever employee, taking this to heart, orchestrated a practical joke on Barbin and his cofounder, Narinder Singh.

"Someone suggested a great idea to me for a joke to play on Narinder," Barbin recalled. "She suggested that we get bobble-heads made of him and give them out to the entire staff. So I paid for four hundred of them to be made, with my own money, and they're not cheap. I thought this was the greatest idea ever. The punch line is that she ran the same play with Narinder, and now everyone has bobbleheads of both of us."

When Phil Libin of Evernote is away from the office, he can still move around the office and talk to people by using a robot that wheels around on a Segway-like balancing system, with a screen for two-way communication.

"When I'm not at the office, I can log in through a browser and drive the robot around," he explained. "It balances on two wheels, and it's six feet tall. I see through its eyes and ears, and it's got a screen, so people can see me. And so you can have casual conversations at someone's desk through the robot. It's got a laser pointer, so you can shoot lasers, which is just good design. You shouldn't build a robot without a laser. It's definitely pretty nerdy. But, you know, we're nerds. That's in the DNA. Humor's in the DNA."

Seth Besmertnik of Conductor also makes himself part of the joke during a weekly gathering of the entire staff.

"Every Friday we have an all-company meeting where we go over everything from the past week," he said. "They last from ten to forty minutes, depending on what happened during that week. We also go through all the new hires, and I embarrass every new employee by asking them to give a speech in front of the company. We'll ask them to say random things, like why'd you join the company, and sometimes I'll ask something funny, like tell everyone about your most embarrassing relative or something. And we also have this big spin-the-wheel, and the 'Conductor of the Week' spins it. Their names are chosen randomly by an automated program. It's a badge of honor, but the point is for everybody to know everyone else better. They spin this big wheel next to my desk, and depending on where they land, they have to do a different thing.

"One might be a talent show, and they have to perform like a little talent show in front of the whole company, right there on the spot. It lightens up the dialogue. As the company gets bigger, it helps everyone get to know people from other departments a little bit more. Otherwise, it's very easy for you to have colleagues who work on the other side of the office, and you never talk to them. There's also a crown on the wheel. You have to wear a crown around the office the whole day. There's Ferris Bueller's Day Off, which means you get to go home early. There's one that's a picture of me with my veins popping out of my head, and then I have to go make this fake angry face in front of them."

Make Them Compete

Laura Ching of TinyPrints.com said her firm likes to hire people who are particularly competitive. And their fun reflects that culture.

"We'll screen for skills, but I'd much rather choose someone who just can't stand to lose," she said. "And so I ask a lot about trying to get at their competitive nature. Have they played competitive sports before? What about a musical instrument? The debate team? How important was that? Our company gets so competitive. We'll have an Iron Chef pumpkin competition in October, when everyone has to make something in the office with pumpkin as an ingredient. People go crazy. They'll bring in a Bunsen burner to make crème brûlée. Or we'll have pumpkin-carving contests. I think it says something about people that go all out about that as well. We like to have fun. I mean, we really have the attitude that if we're going to work a lot of hours, we sure hope our employees will look forward to what they're doing."

Andy Lansing, the CEO of Levy Restaurants, made competition part of his usual visits to locations around the country.

"I'm not a fireside-chat kind of guy, doing company updates that are very formal. So we have a really fun thing called 'On the Road with Andy,' where I'll take my Flipcam with me whenever I go to one of our locations and we'll do a video, either about a great employee we want to highlight or about an incredible food item they're doing at a particular location that I want the rest of the company to see. It's real tongue-in-cheek and fun and we post those for the whole company to see," he said.

"We just did a really neat feature where the whole company participated in a contest. We called it March Madness. Everyone's always saying, 'Andy, we want you to come to our location, we have something to show you.' So all one hundred locations submitted a one-minute video of why I should come to their location, what they want to show us. And the whole company used brackets, like March Madness, with two videos that they were voting on online. One location got some pro athletes to say, 'Come on Andy, you've got to come here and see this.' Some had mascots doing things. To me it's about those kinds of fun, human things that help set the culture.

"I don't like the idea of being a corporate CEO with formal messages. I don't like the town hall where you have to line up with a microphone. It's not who I am. So the more we make it casual and the more we use humor, the better. You don't have to be a comedian, but humor, to me, is the world's best tonic. We work our tails off in the hospitality business, but if you can do it and laugh and have a good time doing it, it's really special."

Cathy Choi of Bulbrite said that the employees didn't have to look very far for a team-building exercise during the holiday season:

"We do a lightbulb decorating contest every Christmas. People pair off in teams of two, and they get half an hour to decorate them, and we hang them up on the tree. We give out awards, like 'Most Likely to Be Recycled,' which basically means the ugliest. It's become very competitive."

At Angie's List, the company uses its parking lot for an annual competition.

"The Friday before Memorial Day weekend, we hold a soapbox derby, and we break the company up into teams from all different departments," Angie Hicks said. "So you get to meet folks that you may not typically work with. And then in the morning, you get to build a soapbox derby car with the seventy dollars we give people to go buy the supplies. Then we race in the afternoon in our parking lot, which has a slight slope to it. Some of our vendors will even fly in to participate."

Richard D. Fain, the CEO of Royal Caribbean Cruises, said that creating debate teams was as fun as it was productive:

"We were having some discussions about prioritizing things, like whether we need better computer systems or better company policies. What we did one year, which was just a tremendous amount of fun, was to make two teams of people, six people on a team from various departments and across our cruise line brands, to lead a debate. And we told each of them to argue a position. And they spent about two months preparing for the debate.

"There were five of us as judges and forty people to watch the debate. It was thrilling, just thrilling. I mean, they were so passionate. And by the way, one of the people was head of sales for Royal Caribbean and argued against what her position would normally have been. The caliber of the work they did was brilliant.

"It was one of the better things we've done in a long time. First of all, I had heard endlessly beforehand why we needed to do each of the things. Here we had the two things competing with one another, and all of a sudden, you saw elements that nobody had raised, so you saw weaknesses on both sides. It transformed the debate. They brought stuff to the fore that I had never remotely imagined was relevant or was happening. We ended up deciding that instead of doing two projects, we would focus on one. And absent this, we absolutely would have done two mediocre projects instead of one transformational project.

"My experience is that people love to be challenged. If the challenge is reasonable, or even slightly unreasonable, they love it and they rise to the occasion. There's just no question. People love to be challenged and they love to show off their skills and talents."

Make an Art of Ad-Libbing

Andrew Cosslett, the former CEO of Intercontinental Hotels Group, shared a memorable story that illustrates the valuable skill of ad-libbing and having some fun on the fly.

"We created this group of the top two hundred job-title holders in the company when I first arrived," he began. "I said, 'Look, we're not going to do what we need to do fast enough if we use the normal chain of command.' I can get to know two hundred people quite quickly because I'm interested in people, and I've got a good memory, so I won't forget who they are or what they do. We set up this group called the Knights of the Round Table. And we've got all the iconography around it, and we have an annual award, a knights award, which is about collaboration—not about achievement, but about whether you've helped others achieve. So we set up this whole code of honor and behaviors and values that we're expecting out of this group."

I told Cosslett that "Knights of the Round Table" sounded cheesy.

"It may sound cheesy, but it works," he said. "I have a very high level of sensitivity to cheesiness, by the way. The knights group sounds that way, but the way we've set it up, they've embraced it and taken it forward. I put a few things out there, and then they have to be adopted. If you're forcing it, it's not going to work.

"When I brought this group of two hundred together, I was looking for a gimmick or something. And it just happened that we were meeting at a hotel in a room called the Courts of King Arthur Room. And it's got knights and suits of armor and old heraldic flags and oak walls, and that's where we happened to have chosen the conference to be. And I walked in and went, 'Oh, geez,' because it was dark. And then I thought, "Why don't I benefit from this? Knights. Knights. Oh, Knights of the Round Table. That's what I could call this group.' And I just sort of invented it on the spot. It came across like it had all been preplanned, of course, which is the serendipity of leadership sometimes. How you ad-lib is very important.

"Because they were coming together as a group for the first time, they liked the identity. I started calling them Sir Leslie and Sir Harold and Sir Adam. We had fun with it, and it started to gain traction. And we put some awards out there, and they're very proud of it. I mean, the fact that we call it knights is neither here nor there."

And here's the key insight from Cosslett for leaders about fun in the workplace: "If you can laugh at yourself, then you're protected either way. It's just getting that balance right. It's the balance between cheesy and fantastic."

16.

ALONE AT THE TOP

As a leader, I never expected people to like me,
but it matters if they trust you and respect you.

—Sir Terry Leahy,
former CEO of Tesco

Steve Case recalled a lesson about the role of a leader that he learned in his mid-twenties from Jim Kimsey, a fellow cofounder of AOL:

"My view, in the early part of my career, was that appearances mattered, that looking like you're working hard mattered. I remember Jim saying this once—and partly I think it relates to some of his training in the military—that really the art is trying to set the priorities and assemble a team so you wake up in the morning and actually have nothing to do. It's impossible to achieve, but it's a good goal to have the right priorities and the right team in place so they can execute against those priorities. It's almost the opposite of how I was approaching it. The objective should not be looking busy, but actually creating a process that allows great things to happen in a way that you can be less involved. So it was sort of a process of letting go, which is hard for entrepreneurs. But at some point you've got to let go and you've got to step back. Ultimately that is about trusting the people you've got but also trusting yourself, that you've set the right context in terms of the vision, the priorities, the team."

It's certainly an enticing notion. What if, as a leader, you were able to develop a sound strategy, and assemble a crack team to execute the plan? What if you were able to establish a culture that addressed many of the points in the earlier chapters—with a simple plan, clear values, a culture of respect, accountability, and candor (but not "candor" via e-mail)? What if you put into practice some of the ideas described here, about building better managers, communicating constantly, surfacing problems, fostering continual learning, running more productive meetings, knocking down silos, innovating effectively, and having some fun once in a while?

If all those goals were achieved, what then would be the work of the leader?

Making Hard Choices

Leaders have to view their staff as a team—and if people aren't playing their position, they have to leave. The decision to let people go can be difficult, but performance matters.

Caryl Stern of the U.S. Fund for UNICEF issued a tough directive to her managers when she first took over.

"As a new CEO, I had breakfast with every single staff person over the course of the first six months," she said. "And then I said to every boss, every senior employee: 'I want your brightest and your best. Give me a list. Who are your brightest and your best?' I didn't tell them how many names. They all gave me their lists, and I said: 'Okay, you've got one year. At the end of the year, either everyone working for you is on this list, or you're telling me how you're getting them there or you're getting rid of them. If we are going to attract the brightest and the best, then we've got to keep only the brightest and the best.'

"That was one thing we changed immediately. Some people were really good at what they did, but they were really difficult to work with. They're all gone. I can teach people skills. I can't teach them how to play in the sandbox."

Julie Greenwald of the Atlantic Records Group takes a similar approach with her top staff to make sure her company isn't growing complacent.

"I constantly ask my senior team, every four to six months: 'Do you have the best people working for you? Do we really have people who are good enough to take your chair? And if not, let's get rid of them. It's okay because there are great new people out there. Let's make sure we always have the best here,'" she said.

"So, I'm constantly asking and pushing my staff to make sure that they feel like they've made the right hiring decisions. And if they didn't, it's okay. It's cool. It's not a reflection of them. Let's learn from it, and now let's get the best person in there. And I try to let everyone know it's about all of us, and that if you do have the wrong person, it's not fair to all of us for you to have that weak link there. I'm constantly pushing and asking that because I don't want complacency."

Geoffrey Canada of the Harlem Children's Zone said he learned a hard lesson that leaders can only do so much in encouraging people to play their position.

"At a school in Massachusetts where I once worked," he said, "we managed early on through consensus. Which sounds wonderful, but it was just a very, very difficult way to manage anything, because convincing everybody to do one particular thing, especially if it was hard, was almost impossible.

"There were about twenty-five teachers and instructors and others. And very quickly I went from being this wonderful person, 'Geoff is just so nice, he's just such a great guy,' to 'I cannot stand that guy. He just thinks he's in charge and he wants to do things his way.' And it was a real eye-opener for me because I was trying to change something that everybody was comfortable with. I don't think we were doing a great job with the kids, and I thought we could perform at a higher level.

"It was my first realization that people liking you, and your being a good manager, sometimes have nothing to do with one another. And I really like people to like me. I was always the kind of person who was a team player. Then I found out that that worked in theory just fine, but it made no sense when you were trying to do difficult things. Managing under those circumstances became difficult because I think it's one thing to manage when people are rooting for you. It's different when people really aren't rooting for you, and they want your plan not to work to prove that they were right and you were wrong.

"Convincing people to give your way a try will work if you neutralize—and sometimes you have to cauterize—the ones who really are against change. They're the kind of person who, if you tell them it's raining outside, they'll fight you tooth and nail. You take them outside in the rain, and they'll say, 'But it wasn't raining five seconds ago.' I spent a year trying to convince those people to change and give me a chance. Then I realized that was a wasted year. I'd have been much better just to simply say, Okay, thank you, difference of opinion. Go do something else with your life. Let me work with this group of folks and move forward. And then you can rebuild that relatively quickly.

"Now I am very clear with people that I will respect your opinion, and I will listen to the range of issues on the table, but once a decision is made, even if you don't agree with it, it is your job to make me right. That's just how it goes. Then, in the end, if it turns out that we've worked as hard as possible and I'm wrong, I'll just say, 'Okay, so let's change.'"

Amy Astley of *Teen Vogue* described how she grew comfortable enforcing standards and delivering tough news.

"In my early job at *Vogue*, and now at *Teen Vogue*, you're managing creative people," she observed. "It's very different from managing people who are doing quantitative work. It's all qualitative, and it's all you judging their work. And it becomes very emotional, particularly when you're judging a writer. I can remember shifting a young person on my team out of an assignment so that I could use a more seasoned writer at *Vogue*. I was trying to make the products that I was responsible for as special as they could be. I left a lot of bruised feelings in my wake. And I really learned from it.

"That said, one thing I did with people is give them a little bit of tough love, too, and say to them, 'We need to make our product the very best it can be. And you need to work with me. You're part of my team. Here are the five great things you did. I'm sorry we have to shift this one to someone else. I think you could learn if you watched what they're going to do with this.' At some point, the person has to see that somebody else is bringing more to the table. It's a bit harsh. But if you work in that environment, you have to be okay with that.

"I think that as you get older, and now I am older, it's much easier. People will accept it much more. But for a young manager it's really difficult, I think, for people to accept your authority. And now I try to be gentle with people, but I also make no apologies because I do feel that when you're running a business, you need to make the hard decisions. Everything has to be the best it can be. If someone can handle the task better than you, it's just got to be that way. I'd want the strongest horse to pull the cart up the hill. That's it. I have a high standard for excellence. I try to have people around me who can accept that that's our goal and that's where we need to be. I'm not great with working with people who are uncomfortable with that."

Setting the Pace

One of the greatest challenges for leaders is to strike the right balance in creating a sense of urgency and change. Too much and too fast, and people will be left behind. Too slow, and the competition will jump out front. Many leaders talked about the importance of consistency, smoothing out the highs and lows, setting the right pace for change.

"Pacing is really important in an organization," said Harry West of Continuum. "When you're leading, you're generally trying to lead change, and I think it was [the scientist] Roy Amara, who said about technology, 'We tend to overestimate the effect of a technology in the short run and underestimate the effect in the long run.' And I think the same applies to change within an organization.

"I have in the past tended to overestimate the amount of change I can effect in the short run and then not fully appreciate the change I can effect in the long run. And so I've learned that it's critical to think carefully about the pace of change, and it's something that I've learned the hard way. It's important to manage that carefully, because it's not just about the pace of change that certain people in the company can manage. It's about the pace of change that the company as a whole can manage. You can push and push and nothing seems to happen, and then suddenly it takes off and you're sort of running to catch up."

Carl Bass of Autodesk captured West's insight in a memorable image about companies and speed.

"In a small company, most of the people are aligned. You can wake up any day and head in a different direction," he said. "On the other hand, a bigger company takes more work and takes more steering. And the analogy for me is that a small company is like a ball-peen hammer; you can move it back and forth really quickly.

Big companies are more like sledgehammers. It takes a lot to move them, but when you do you can actually have a big impact. So you start realizing that a lot of the work you're doing, which might not feel that satisfying in the short term, is really necessary so that you can swing the sledgehammer and have a bigger impact.

"Another thing I've thought about as I've run a larger organization is that, as CEO, you're the one who's driving the bus. And if you're erratic while you're driving, everyone gets pretty nauseous. It's really important to be as clear as you possibly can be and not just wake up one day and say we're going this way and the next day we're going that way."

Dennis Crowley of Foursquare said he learned to make sure he keeps his staff focused on the current plans, even though he is already seeing several steps down the road.

"I've learned to have more discipline in the way we talk about certain things," he said. "I keep a notebook in my pocket and I write down all the stuff we could ever do with Foursquare. I used to go through the process of sharing that with everyone. 'As soon as we finish this, we're going to go on to this, and then we're going to go on to this . . .' I get really excited talking about that stuff, but for people whose job it is to execute on the current plan, I've learned that it can be distracting—as in, 'Whoa, I'm supposed to be doing this, but the CEO is really excited about this. Should I do this instead?' So I've learned when to bite my tongue about things I'm excited about."

Rob Murray of iProspect tells his staff to expect constant change in their roles and the structure of the company. That way, they're not surprised when it happens.

"We're evolving really fast and I've always really believed and pushed hard that you have to embrace change and be willing to evolve," he said. "We've made major structural changes to our company about every eighteen months. You will not survive and aspire to greatness if you assume there's one way of doing things and just continually beat that drum over time. So we have to have a team that's very flexible and adaptive, and understand that change is going to happen and we're going to do things a certain way today, but then let's assume in eighteen months we're going to have to change it."

Katherine Hays, the CEO of GenArts, a visual-effects technology company, is one of several leaders who discussed the importance of smoothing out the roller coaster of business.

"It's important to keep things in context, whether it's good news or bad news," she said. "Either can be very distracting to the team. I'm pretty good at keeping those in context and focusing on the task at hand. Some of the boards I've worked with are really good at that as well. They just don't overreact, no matter what the news is.

"I learned as an athlete—I rowed for four years in college—that you have to be present in the moment, and you can't be distracted by something you just did that was really good, or by the fact that you're a little bit behind in a race. You can't focus on what's just happened, because you can't change it. That's not to say we shouldn't pause and congratulate ourselves, but you have to balance that with maintaining focus on what the next steps are. You learn as an athlete to say, 'Great, we won that race, but what are the things we could have done better? Because we have a race next week.'"

Joel Babbit of Mother Nature Network learned over time to take problems more in stride.

"Early on, I was much more emotional on every level. I would make decisions more emotionally. I would confront problems more emotionally. I would react to huge successes more emotionally," he said. "And I certainly have found to some degree that it's not the end of the world when something goes bad, and that that's going to happen a lot. And when things go well, I've learned to not overcelebrate because it's probably not as good as it looks. So over time I've just learned that the highs and lows are never as high or low as you think they are. I haven't learned that lesson one hundred percent, however. I still get crazy every once in a while."

Steve Case said that he learned to play the role of "shock absorber" as the leader of AOL when it was growing quickly.

"I was pretty calm, a cool operator," he said. "Even my nickname internally was 'The Wall,' which is a little pejorative, and probably that's because there's not as much connection with people or empathy as I would probably like. But the other side of it was that we were dealing with a lot of challenges and had a big idea. We really were trying to change the world, and there were enormous ups and downs. One day you're the smartest company in the world, and you're going to take over the world, and the next day you're the dumbest company in the world and you're going out of business.

"These fluctuations happen constantly over a ten-to-fifteen-year period. So I adopted the role of being the company's 'shock absorber,' particularly with the management team. My basic view was, we're never as good as people say we are when times are good and we're never as bad as people say we are when times are bad. What I'm going to try to do is even out those enormously volatile swings. As a result, when times were good I would tamp down on celebration—some was okay, too much was not okay—and also deliberately delegate paranoia.

"Basically, that meant getting other people out of their comfort zone. If they felt like it was time to have a victory parade, I'd say, 'Well, it's great, but what about this and what about that? Don't spend so much time celebrating. We've got to get ready for the next battle, the next challenge.' Conversely, when the chips were down and people were losing confidence and losing hope, I would bring people back up and remind them why we're here, why this is a battle worth fighting and why it's an important battle and why we're well positioned to win.

"I guess there are some companies that are overnight successes, but they're few. Most are pretty choppy, and I've found that it's important to recognize that and put it in a context where there's a little bit more balance."

Building Trust

Effective cultures are built on trust—among colleagues, and in the leadership of the organization. For employees to bring their "best selves" to work, they have to feel confident that all the rules of the organization, stated and unstated, will hold, and that the leaders will remain true to their word.

Employees understand that bosses will make mistakes, and that they may have to change course suddenly because a strategic bet didn't pay off. But they have to sense that the leader is trustworthy, has integrity, and respects the people who work for him or her—and that every decision, even if it's not popular, is at least made in the best interest of the organization.

John Riccitiello, the former CEO of the videogame maker Electronic Arts, was particularly thoughtful on how trust is essential for effective leadership.

"One of the things I would say about leadership is that you have to be absolutely genuine," he said. "You have to know what you truly believe and what you truly value, and it has to be undeniably consistent. Many years ago, we were in this interesting spot where our traditional business was in trouble. I could clearly see this digital transformation around social networks and mobile phones. You don't know exactly how you're going to make money on the other side. But if you stop being consistent, then nobody has the confidence to go along.

"So while we were going through this radical transformation as a company, everyone could count on two things: You could be sure that while we were cutting, we were never going to sacrifice the quality of our product. And the second thing is that if you were a key contributor to a process of bringing a great product to market, not only were we going to support you, but my number one job is to get the blockers out of the way so your product can find a marketplace. Those were the two things that were consistent. Everything else changed. If you're going to ask people to go along with you, when almost everything they know about their job, their company—how it makes money, how it works, how Wall Street is going to view it—is going to change, you've got to pick a couple of things and stay with them. We had to have something that was foundational.

"And so everyone knows what I stand for. They're not going to follow you if they don't know what you stand for. These same things need to be what you're willing to fire somebody for. You can't have cynicism. And what causes cynicism is when you allow executives to live in an organization and prosper even though everything about them undermines where you're trying to go as a company.

"Of course, you don't always win with strategic bets. But I wanted to make sure that on the other side of the things I was asking people to do, people I respected would still respect me, the people I liked would still like me, and they were more likely to be friends at the end than not. I wanted to be that person on the other side, so I needed to define something that I could stick to. I want to respect myself in the morning, and I think a lot of it started with that—what would allow me to do these really hard things that meant that some people were going to lose their jobs, and that almost everyone was going to doubt. I needed a foundation stone for myself."

Sir Terry Leahy, the former CEO of the British grocery company Tesco, said that trust became a cornerstone of Tesco's culture after he led a process of codifying the company's values.

"Just before I became CEO," he said, "I brought all of the Tesco people together in small groups—it took over a year—and I asked just two questions. 'What do you think Tesco stands for? And what would you like it to stand for?' That was the more revealing question because of the golden rule of treat people the way you'd like to be treated. They were prepared to dedicate themselves in the business to service, but they wanted a culture that was respectful and provided dignity for people. It was amazing how simple it was and how it coalesced around these two pillars of service and good manners.

"If I had to sum it up, it would be about being generous at work rather than selfish. And it is amazing how often you see people who can't help themselves, because of their ambition or their insecurities or whatever—that they're basically selfish and they take rather than give. For some people, that's a transition that they have to make, and not everybody can make it. Sometimes the brightest find it the hardest to make that transition because they've always been better than the people around them. They find it hard to trust the people around them to do the work. They think, 'Well, I know best.' And when you see organizations that struggle, it's mainly that people can't trust. The leaders can't trust, and then the teams don't trust each other. You have to create conditions where people can work together because they trust each other, and that really empowers the organization.

"It also has a lot to do with a process of making people feel good about themselves. Bureaucracy in organizations will tend to lower self-esteem, and so if you consciously build people up so they say, 'I matter here and people respect me and they think I can contribute and they trust me to contribute,' that really gets the best out of people. As a leader, I never expected people to like me, but it matters if they trust you and respect you. I think that's mainly around consistent behavior. If you treat one person differently from another, or if one day you're different than the day before, then it's really hard, because they don't know where you're coming from. People have got to know where you are coming from. And you have to be a winner. It's hard to maintain trust in someone if they ultimately are not successful. You have to get enough right."

And that is why it is lonely at the top. Although leaders in many organizations can seek advice from their boards of directors and executive teams on key decisions about strategy, there are many moments when leaders have no one to consult but themselves for tough calls—ensuring only the best employees are on the team, setting the right pace for their organization, and making decisions that inspire respect, confidence, and trust. Such moments can be overwhelming for some, while others are drawn to them.

"If I'm in a room or a crowd of people, I tend to just get involved," said Arkadi Kuhlmann of ING Direct. "Usually, I really like whatever the problem is. I like to get close to the fire. Some people have a desire for that, I've noticed, and some people don't. I just naturally gravitate to the fire. So I think that's a characteristic that you have, that's in your DNA."

CONCLUSION

Leadership is endlessly challenging for many reasons, including this one: leaders must be comfortable with seeming paradoxes, an ability to find the balance point between two opposing forces and impulses. This balancing act can exist, for example, in the values of an organization.

"We are building companies, and so we have to be really accountable," said Jacqueline Novogratz, the CEO of the Acumen Fund, a global venture fund that uses entrepreneurial approaches to help poverty-stricken regions. "We've got to be tough, and yet we have to be very generous, since we're working in communities where people make a dollar or two dollars a day. We talk about the power of listening and we juxtapose it with leadership, because sometimes you've listened enough, and now it's time to make a decision.

"We think about our values in pairs, and there is a tension or a balance between them. We talk about listening and leadership;

accountability and generosity; humility and audacity. You've got to have the humility to see the world as it is—and in our world, working with poor communities, that's not easy to do—but have the audacity to know why you are trying to make it be different, to imagine the way it could be."

This is a useful framework for thinking about fostering a culture of innovation, because that challenge also represents a paradox. Leaders, after all, must reduce the chaos inside their organizations, making predictable rules and behaving in predictable ways, so that employees can better respond to the swirling changes of the marketplace outside their organizations. With a consistent (but not overly rigid) approach to "how"—including how employees are expected to treat one another and work together—leaders can more effectively focus their teams' attention on the work at hand.

"I believe most companies fail because they're not focused," said Ryan Smith of Qualtrics. "They either get focused on other things in the market that aren't important, so they're thrashing around without a clear objective, or they're focused internally on things like politics and bureaucracy. It's not that these companies aren't smart companies and they don't have good businesses. It's just that there's a lot of noise."

My goal in writing this book has been to help leaders reduce the distracting noise in their organizations so that they can focus on building a high-performing culture. I also hope that leaders will find this book helpful for managing their time effectively, since that is one of the biggest challenges they face, given the endless demands on their schedules. On the daily to-do list, there are inevitably more urgent matters—fires to put out, reports to file, personnel problems to solve, dozens of e-mails to write—than tending to matters of their organization's culture. After all, the return on investment of their time is much clearer with more tangible work. When it comes time to focus on culture, where to begin? What to do? Where to start? It can be easier not to deal with culture at all, given the pressure to deliver immediate results.

But as leaders sit down to allocate their time and assess whether they are using it wisely, this book can serve as a guide, so that they will be confident that time spent on working with employees to codify the values of their organization—a potentially disruptive process that may result in some people leaving the company, by choice or not—will ultimately build a stronger culture. When leaders bring their staffs together for a weekly or monthly all-hands meeting, and wonder whether there is any benefit in repeating for the umpteenth time the broader strategic goals of the organization, they will understand the payoffs of regular and consistent communication. When they wander down the hall to have a face-to-face conversation about a sensitive subject, rather than shooting over a quick e-mail about it, they will know that they have spent their time well.

No solid line connects any of these strategies to a clear impact on the bottom line. But there is a powerful case to be made, as the CEOs in this book have shown, that in the long run, culture can be a secret weapon as leaders build their organizations. If they can bring out the best in their employees, and create a culture of innovation, then they will foster better performance and the kind of creative thinking that companies need to survive and thrive.

That also requires embracing, and pursuing as a goal, the paradox captured in the first words of this book, from Dominic Orr of Aruba Networks: "We aspire to be the largest small company in our space."

To Orr, a high-performing culture is the key to winning. "The only reason we have come so far and we'll be able to sustain our gains in market share is because we have more focus and we move faster," he explained. "I tell everybody, 'If you think we have better talent than our competitors, dream on, because they have very deep talent, too.' The only thing that we can do is we can focus and we can move faster.

"And the one thing that allows me to move faster than the much bigger companies I compete with is I have less politics. I don't want politicking, and I'm truly convinced that politics arise because people dig into their position because they have ego tied to it. I'm trying to extend my speed as long as possible, and I defuse a lot of politics by telling people, 'You don't need to dig into your position. Just be intellectually honest.' That's my tool to break up all these potential blocks of ice that then maybe become icebergs."

Simple rules like Orr's can have a powerful effect on a corporate culture if employees trust that those rules will be applied uniformly and consistently across the organization. Sometimes just two values—like the ones at LivePerson: "Be an owner" and "Help others"—can set the tone for an entire company. Such values can serve the role of rules in a sport, like football. If they're lacking, or if they're unevenly enforced through the equivalent of bad refereeing, players start throwing up their hands and shaking their heads in dismay; they may resort to dirty tricks to get an edge. They may want simply to play a good game of football, but a culture that evolves by default, rather than by design, can bring out the worst in the players. This effect is magnified in the workplace because so much is at stake—ego, identity, a paycheck, and the prospect of future promotions.

Shifting this dynamic represents an enormous opportunity for leaders. By focusing more on culture, they can achieve a huge lift as people bring their best selves to work. And most people do want to be led, they do want to pull together as a team, they want to believe in and achieve something greater than themselves.

"We preach a lot here that team is one of the most beautiful of all human experiences," said Kip Tindell of The Container Store. "You do great things together, and you go home at night feeling wonderful about what great things you accomplished that day. That's what people want, and that's what wise and sophisticated leaders help cultivate and know that people want. Every bad boss you or I have ever had thinks that what people want is the exact opposite of that."

The value in marshaling a sense of teamwork cannot be quantified on spreadsheets, captured in flow charts, or circled as an entry on a balance sheet or quarterly report. But as we move into an increasingly knowledge-based economy, there are tremendous gains to be achieved in finding ways to encourage people to work together more effectively.

"Especially nowadays, you're hiring individuals to think," said Ryan Smith. "We can't control the way they think. All we can control or have an effect on is the environment around them."

The companies that will thrive over the long haul will understand that culture is a key element of their strategy—for attracting and retaining the best talent, for encouraging employees to bring their best selves to work, and for fostering an environment in which everyone feels motivated to innovate. As the global economy presents more challenges to leaders, those who create a quick and nimble culture will emerge as winners.

ACKNOWLEDGMENTS

"Corner Office" started with a simple idea: What if I sat down with CEOs and never asked them any of the usual business questions about their strategies and competitive landscapes, and instead asked them about leadership lessons they've learned, the culture they try to foster at their companies, and how they hire?

That initial idea has grown into much more than I ever expected, and I am deeply grateful to the team that helped make it happen. The *New York Times* provides a powerful stage for my "Corner Office" columns and for this book, and many colleagues in the newsroom championed the idea from the start. Jill Abramson, the executive editor; Dean Baquet, the managing editor; and Janet Elder, an assistant managing editor, have encouraged me to take the feature in new and bigger directions. I have relied on the guidance of Larry Ingrassia, an assistant managing editor who was business editor when we launched "Corner Office" in 2009, as well as that of Dean Murphy, the current business editor; David Gillen, the section's enterprise editor; and Vera Titunik, the Sunday Business editor. Rick Berke has been a big supporter of my work on "Corner Office," and I have learned a tremendous amount about leadership from him as his deputy on the national desk and the features sections. Special thanks to two editors in the Sunday Business section—Phyllis Korkki, who edits the column each week, and Dan Cooreman, who has a deft touch with headlines that capture the essence of each column.

This is my second book with Times Books, and I have been lucky to work again with the publishing team of Paul Golob, a masterful editor whose wise suggestions always lift a manuscript, and Steve Rubin, Maggie Richards, Pat Eisemann, and Emi Ikkanda. Alex Ward, the editorial director of book development at the *New York Times*, lent his usual steady hand to this project, and I relied on the savvy advice of my agent, Christy Fletcher, and her team at Fletcher & Company.

My father, Clellen Bryant, a longtime editor at *Time* magazine and *Reader's Digest*, reads the first drafts of my books before anyone else. He has a remarkable eye for nuance and shadings, and I have tried to emulate his skills—not just with words, but also for bringing out the best in writers—in my own editing roles at *Newsweek* and the *New York Times*. Thanks to my mother, Julie, for her constant support. And thanks to my stepmother, Jill, for giving the book such a close read.

It has been fun to share what I've learned from the leaders I've interviewed with my daughters, Anna and Sophie, as they attend college and embark on their careers. It is a key milestone of life when you realize your children are smarter than you about a lot of things, and that list is no doubt going to grow longer with each year. Though I'm naturally competitive, this game is one I am happy to lose.

Writing a book can feel at times like scaling a mountain, with a little high-wire walking between outcroppings just to add to the challenge. Though a project like this is a mostly solitary affair, my wife, Jeanetta, has been my constant companion, providing guidance and the benefit of her pitch-perfect instincts along the way. I could not have done this without her.

INDEX

ABOUT THE AUTHOR

ADAM BRYANT is the author of the *New York Times* bestseller *The Corner Office: Indispensable and Unexpected Lessons from CEOs on How to Lead and Succeed*. He writes the popular "Corner Office" feature in *The New York Times* and has served as the newspaper's senior editor for features, its deputy national editor, and its deputy business editor. He was previously a senior writer and business editor at *Newsweek*. He and his family live in New York City.